Teubner Skripten zur
Numerik

Peter Oswald
Multilevel Finite Element
Approximation

Teubner Skripten zur Numerik

Herausgegeben von
Prof. Dr. rer. nat. Hans Georg Bock, Universität Heidelberg
Prof. Dr. rer. nat. Wolfgang Hackbusch, Universität Kiel
Prof. Dr. phil. nat. Rolf Rannacher, Universität Heidelberg

Die Reihe soll ein Forum für Einzel- sowie Sammelbeiträge zu aktuellen Themen der Numerischen Mathematik und ihrer Anwendungen in Naturwissenschaften und Technik sein. Das Programm der Reihe reicht von der Behandlung klassischer Themen aus neuen Blickwinkeln bis hin zur Beschreibung neuartiger noch nicht etablierter Verfahrensansätze. Es umfaßt insbesondere die mathematische Fundierung moderner numerischer Methoden sowie deren Aufbereitung für praxisrelevante Anwendungen. Dabei wird bewußt eine gewisse Vorläufigkeit und Unvollständigkeit der Stoffauswahl und Darstellung in Kauf genommen, um den Leser schnell mit aktuellen Entwicklungen auf dem Gebiet der Numerik vertraut zu machen. Dadurch soll in den Texten die Lebendigkeit und Originalität von Vorlesungen und Forschungsseminaren erhalten bleiben. Hauptziel ist es, in knapper aber fundierter Weise über aktuelle Entwicklungen zu informieren und damit weitergehende Studien anzuregen und zu erleichtern.

Multilevel Finite Element Approximation

Theory and Applications

Von Prof. Dr. rer. nat. Peter Oswald
Universität Jena

B. G. Teubner Stuttgart 1994

Prof. Dr. rer. nat. Peter Oswald

Geboren 1951 in Dresden. Von 1970 bis 1975 Studium der Mathematik an der Staatlichen Universität Odessa (Ukraine/Sowjetunion), 1975 Diplom. 1978 Promotion am Moskauer Institut für Elektronischen Maschinenbau und 1982 Habilitation im Fach Mathematik an der Friedrich-Schiller-Universität Jena. Von 1978 bis 1988 Assistent, Dozent an der TU Dresden. 1981/82 Zusatzstudium an der Moskauer Universität. Seit 1988 Professor für Numerische Mathematik an der FSU Jena.

Die Deutsche Bibliothek – CIP-Einheitsaufnahme

Oswald, Peter:
Multilevel finite element approximation : theory and applications /
von Peter Oswald. – Stuttgart : Teubner, 1994
   (Teubner Skripten zur Numerik)
   ISBN 978-3-519-02719-5   ISBN 978-3-322-91215-2 (eBook)
   DOI 10.1007/978-3-322-91215-2

Herstellung: Druckhaus Beltz, Hemsbach/Bergstraße

Preface

These notes reflect, to a great part, the present research interests of the author but were influenced by the ideas and the work of many colleagues. They are based on lectures given by the author at the Institutes of Mathematics and Informatics at the Technical University of Munich during February/March 1993. I wish to warmly thank Chr. Zenger and R. Hoppe for their generous support and the many discussions I had with them and their younger colleagues during the last year. Part of the results contained in section 4 is the output of these discussions and joint work with M. Griebel.

There are many other mathematicians who encouraged me (or personally or by their mathematical work) to step into the field of multilevel methods. I want to acknowledge the support I received from W. Dahmen, R. A. DeVore, P. Deuflhard, W. Hackbusch, H. Triebel, O. Widlund, H. Yserentant and many others. On the other hand, I should apologize for not mentioning many interesting research results and names standing for recent developments in the fields which are the subject of these notes.

Finally, I want to thank my family, my wife Olga and my daughters Evelyn and Annelie, for their everyday patience and support.

Dresden, May 1994   Peter Oswald

Contents

8   Contents

1　Introduction

These notes are an attempt to collect some specific information from approximation and function space theory and to present it in a form understandable for specialists working on large scale computational methods for partial differential equations. Though theoretical results on approximation processes and the use of various types of function spaces are well-recognized as very important in numerical analysis, recent developments in the field of multilevel-multigrid methods as well as the introduction of the wavelet concept have shed new light on the ties between these mathematical disciplines.

Our main aim is to survey the approximation-theoretical background of what we call stable splittings of Sobolev spaces with respect to multilevel finite element schemes. We can rely on the many recent results on the decomposition method (especially with respect to various types of locally supported functions) which is an important technique in function space theory and Fourier analysis. Such multilevel splittings serve as the basis for designing quite optimal practical solution procedures for a number of p.d.e. problem classes. E.g., modern iterative methods such as multigrid and domain decomposition schemes, and adaptivity concepts can be dealt with.

These are research notes on the finite element multilevel method rather than a textbook or a monograph. The main reason is that the subject has been developing very rapidly over the past few years and is still not finished in many respects. Moreover, we did not feel competent enough to describe all theoretical and algorithmical aspects with the necessary mathematical depth. In many places, the reader is referred to the original literature for full proofs and more detailed information. Often we formulate and discuss problems rather than solving them. What we hope, however, is that the reader will obtain a more precise understanding of the mathematical background and the computational potential of the multilevel method, and will be better prepared to thoroughly solve the open problems.

A few comments on the prerequisites that we expect from the reader are necessary before we start. We assume some knowledge about theoretical and numerical methods for p.d.e.'s in general and on the mathematics of the finite element method in particular (books like Dautray/Lions [DL], the well-known Ciarlet [Ci] or the more introductory textbooks by Marchuk [Mk], Hackbusch [Ha1], Braess [Br1], Johnson [Jo], Carey/Oden [CO] are quite enough to serve as reference. Also, the language of functional analysis (Hilbert spaces, norms, functionals, ...) and the theory of Lebesgue measure (basics of L_p spaces) will be used occasionally, see the books of Kolmogorov/Fomin [KF] or Alt [Al] (you may actually consult any of your favoured textbooks in this respect).

The situation with textbooks for modern multivariate approximation theory is much worse. The basic concepts and a rather complete picture of the one-dimensional case (functions on $\mathbf{R}$ or on an interval) can be found in the recent monograph by Lorentz/DeVore [LD]. The Russian classics are Achieser [Ah] and, more important for our purposes, Timan [Ti] but they do not contain piecewise polynomial approximation. The monographs by Butzer et. al. [BB, BN] reflect the more functional-analytic viewpoint on approximation theory. Schumaker [Su1] discusses one-dimensional spline approximation in detail, see also de Boor [Bo1], Nürnberger [Nb].

Multivariate splines are the subject of Chui [Ch1] and deBoor/Höllig/Riemenschneider [BHR]. Below we will give more specific references (also to the wavelet and theoretical CAGD literature where approximation problems are dealt with, among others).

Now to function spaces - we use the notation $W_p^m(\Omega)$ (m - integer, $1 \leq p \leq \infty$) from the very beginning. It stands for the Sobolev space of all (equivalence classes of) L_p-integrable functions for which all generalized partial derivatives of order $\leq m$ exist and belong to $L_p(\Omega)$. A short introduction is contained in [Tr1], chapter I.3-5, where also the definitions for $C^m(\bar{\Omega})$, $L_p(\Omega)$ may be found. A norm resp. semi-norm which we frequently use is given by

$$\|f\|_{W_p^m} = \|f\|_p + |f|_{m,p} \, , \quad |f|_{m,p} = \sum_{m_1+...+m_d=m} \|\frac{\partial^m f}{\partial^{m_1} x_1 \ldots \partial^{m_d} x_d}\|_p \, ,$$

where $\|f\|_p \equiv \|f\|_{L_p(\Omega)}$. If not stated otherwise, $\Omega \subset \mathbf{R}^d$ is any open domain, our main interest concentrates, however, on polyhedral domains in $\mathbf{R}^d$. We will often omit Ω in the notations to get them more compact. Standard sources on Sobolev spaces are Adams [Ad], Maz'ja [Mz], Grisvard [Gv]. We hope that the reader will accept (without additional explanations) the few properties of functions from W_p^m that we will use below. Other function spaces are introduced in subsection 2.5 and in section 3. They fall into the scale of Besov-Sobolev spaces, a modern exposition of which is given in Triebel [Tr2, Tr3, Tr4], especially [Tr4] may be studied to catch the main points. We want to mention the books by Nikol'skij [Ni] who actually introduced approximation methods as a basic tool to the study of function spaces, Besov/Il'in/Nikol'skij [BIN] (integral representations for general domains, not recommended as first reading on the subject), see also the survey [BKL], Stein/Weiss [St, SW] (maximal functions), Bergh/Löfström [BL] (interpolation theory).

Last but not least, there is extensive literature on theoretical numerical analysis for p.d.e.'s. Besides the books already mentioned, we recommend the monographs by Girault/Raviart [GiR] (finite element theory of flow problems), Brezzi/Fortin [BF] (nonconforming, mixed, and hybrid finite element discretizations), and Křížek/Neittaanmäki [KN]. Iterative methods (one subclass of which are the multilevel schemes discussed in these notes) for solving large linear systems arising from finite element discretizations are investigated in [Ha2, AB, Dy]. We do not know of a reasonable textbook on adaptivity as this is still a field of intensive research activities with many deep and interesting questions left open.

The contents of the remaining sections can be briefly described as follows. In sections 2 and 3 we collect essentially known material on L_p-approximation by finite element functions, splines, and wavelets, and on related function spaces. We introduce a scale of approximation spaces $A_{p,q}^s(\{V_j\})$ with respect to an increasing sequence of approximating subspaces

$$V_0 \subset V_1 \subset \ldots \subset V_j \subset \ldots \subset L_\infty(\Omega)$$

which turns out to be important for the treatment of multilevel methods.

Section 4 surveys the recent results on iterative solvers for discretizations of elliptic problems. We concentrate on so-called subspace correction schemes which are based on appropriate subspace splittings. The important tools of our approach are the theory of abstract Schwarz methods and the stable splittings of Sobolev spaces $H^s(\Omega)$ into low-dimensional subspaces derived from the results of the previous sections.

In the final section 5 we discuss various mathematical aspects of adaptivity within a multilevel scheme, and relate this problem to investigations in nonlinear approximation theory.

2   Finite element approximation

In this section we study certain linear classes V of approximating functions resp. sequences of them $\{V_j\,,\,j = 0,1,\ldots\}$ which are particularly useful for efficient numerical methods in different fields. They share some properties which can be described in short as follows:

- existence of a well-localized and stable (with respect to L_p-norms) algebraic basis,

- simple recursions (prolongation and restriction operators) for exchanging information between different V_j,

- good approximation properties for smooth functions, e.g., from Sobolev spaces.

Typical examples are given in subsection 2.1. In subsections 2.2-2.4 we study finite element subspaces with respect to their approximation power (direct and inverse theorems for finite element best approximation). The tools to measure smoothness of L_p-functions on a fine scale (moduli of smoothness resp. K-functionals) are explained in 2.2. Some additional information and references on the approximation theory by splines, wavelets etc. is given in 2.5. In the concluding subsection 2.6 we introduce the scale of Besov spaces which in turn can be characterized as approximation spaces (more background is provided in 3.4). This scale will be used throughout the following discussion.

2.1   Finite elements, splines, wavelets

2.1.1   Finite element subspaces

We assume familiarity with the finite element language from [Ci] (small alterations of the terminology may occur). Let Ω be an open, bounded subset of $\mathbf{R}^d$ which is fixed throughout the following considerations. To avoid any discussions on boundary resolution, suppose that the domain is polyhedral, i.e. representable by a finite partition into non-degenerating simplices. It is clear from the context how the properties discussed below can be extended to infinite partitions of unbounded polyhedral domains (e.g. of the whole $\mathbf{R}^d$).

Our examples mainly concern simplicial (triangular in $\mathbf{R}^2$ resp. tetrahedral in $\mathbf{R}^3$) or rectangular partitions of Ω (if Ω itself is composed from $\mathbf{R}^d$-rectangles) although most of the theory works for other polyhedral partitions as well. Partitions (denoted by T resp. T_j) are generally supposed to be *regular* in the following sense: The closures of any two different polyhedral subdomains are or disjoint or intersect in a common vertex, edge, face etc. (no mixtures allowed), and the ratios of the sizes of inscribed and circumscribed balls remain bounded from below by a constant $\gamma_0 > 0$ independently of the subdomain (and of j). Consequently, to each subdomain K of T there may be assigned the diameter $h(K) = \mathrm{diam}\, K$ characterizing its typical size. Compare Figure 1 for examples of nonregularity.

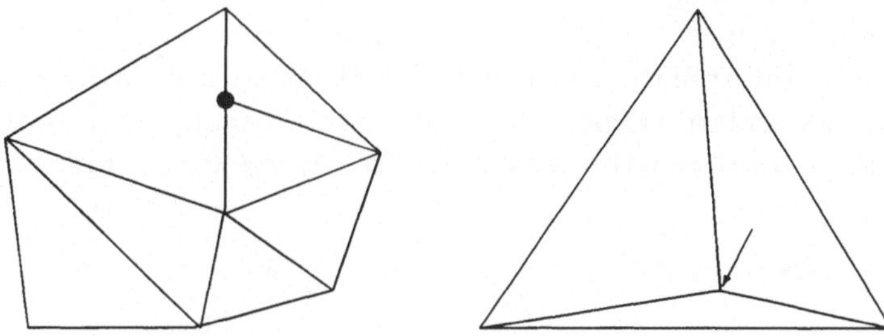

Figure 1. Nonregularity: Slave nodes and degenerating triangles

Usually, $h(T) = \max h(K)$ is called (maximal) mesh-size of T. Let $h_{\min}(T) = \min h(K)$. *Quasi-uniformity* of a partition (or better: of a sequence of partitions $\{T_j\}$) means that the ratio $h(T_j)/h_{\min}(T_j)$ is bounded from above by a constant $\gamma_1 < \infty$ (independently of j). If not stated otherwise, these constants enter the constants appearing in the inequalities below. We will not indicate this dependency in the sequel.

We consider only scalar finite element (f.e.) constructions, vector-valued cases (e.g., for dealing with elliptic systems or for mixed formulations, see 4.6.2) need further elaboration, sometimes they may be reduced to the scalar case. The common point for constructing an f.e. subspace over a fixed partition T is a local procedure: to any subdomain we assign a properly posed interpolation problem with respect to some fixed space of algebraic polynomials. The problems should fit together to yield a globally smooth function belonging to C^0 or C^1 or some other smoothness class. The interpolation problems may be posed in terms of pointwise function or derivative evaluations, equally allowed are functionals using integral means etc. (in general, unique solvability of the interpolation problem in the given polynomial space and compatibility between the different interpolation problems are required). To fit the polynomial pieces together, some of the interpolation conditions, having support on a vertex, edge, face etc. common to two or more subdomains are used repeatedly (for each subdomain the same value will be prescribed!). The f.e. subspace can formally be viewed as the image of the global interpolation problem (composed of local interpolation problems fitted together) into a certain piecewise polynomial space with respect to the underlying partition. The notation $V = V(\mathcal{T})$ resp. $V_j = V_j(\mathcal{T})$ will be used assuming that the context (the family of local interpolation problems) is clear. As the absolute minimum, we have $V(\mathcal{T}) \in L_\infty(\Omega)$ since the pieces are polynomials.

Most of the practically important f.e. constructions use the affine-invariant setting described in [Ci] where all subdomains are of the same type, and the corresponding interpolation problems are obtained from a fixed interpolation problem in a standard element K by affine-linear mapping. Together with the assumption of regularity this automatically ensures some uniform behaviour of

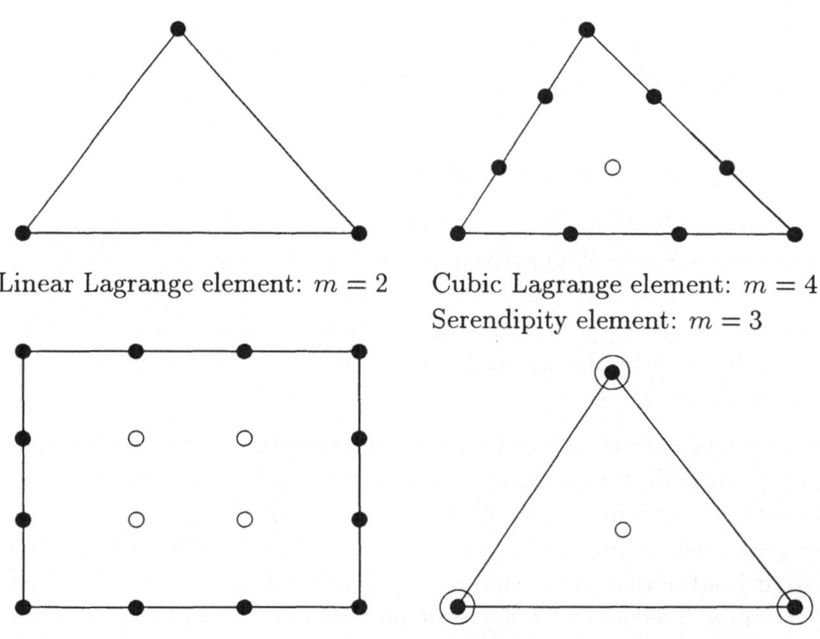

Linear Lagrange element: $m = 2$

Cubic Lagrange element: $m = 4$
Serendipity element: $m = 3$

Bicubic Lagrange element: $m = 4$
Serendipity element: $m = 4$

Cubic Hermite element: $m = 4$
Serendipity element: $m = 3$

Figure 2. C^0 finite elements

the construction over the whole domain. Look at the examples of C^0 elements for second order elliptic equations indicated in Figure 2, more affine-invariant f.e. families are described, together with the technical details, in [Ci, GiR, Br1, BF]. See also the sections 4.3 and 4.6.1 where some examples of C^1 resp. nonconforming elements are given. The description of the different types of interpolation conditions by graphical symbols follows the usual conventions (compare [Ci]): dots • stand for a function evaluation, the symbol o indicates that in the case of serendipity elements no Lagrange interpolation condition will occur at such a point. One or two arrows will be used for first or second order directional derivatives, one (or two) larger circles around a point indicate that full sets of first (and second) order partial derivatives are prescribed at this point etc.

We state now the assumptions we will make use of in what follows (they are obvious for the above-mentioned examples of affine-invariant f.e. constructions on partitions satisfying the regularity assumption), further requirements will follow later.

- **(A1)** There is an algebraic basis $\{N_i\}$ of $V(\mathcal{T})$ consisting of locally supported functions (the suitable normalized nodal basis functions) satisfying the following L_p stability condition: there are absolute constants $0 < c < C < \infty$ such that for all $1 \leq p \leq \infty$ and all f.e. functions

$$g = \sum_i c_i N_i \in V(\mathcal{T}),$$

  we have

$$c\|\{\bar{h}_i^{d/p} c_i\}\|_{l_p} \leq \|g\|_p \leq C\|\{\bar{h}_i^{d/p} c_i\}\|_{l_p}.$$

  Here, $\bar{h}_i$ denotes the diameter of the support of N_i which is characterized by the diameters of subdomains near the corresponding nodal point.

- **(A2)** For some natural m, depending on the f.e. scheme considered, there exists a linear mapping

$$\mathcal{Q} : L_p(\Omega) \longmapsto V(\mathcal{T}) \qquad (1 \leq p \leq \infty)$$

  called quasi-interpolant of polynomial order m satisfying the property

$$\mathcal{Q}p = p \qquad \forall p \in \mathbf{P}_m$$

(with $\mathbf{P}_m$ the class of all multivariate polynomials of total degree $< m$), and such that for two absolute constants c, C and arbitrary $f \in L_p(\Omega)$ we have

$$\|\mathcal{Q}f\|_{L_p(K)} \leq C\|f\|_{L_p(\tilde{K})} \qquad \forall K \in \mathcal{T} \ .$$

Here, $\tilde{K}$ denotes the union of K with all its neighbouring subdomains which are at a distance $\leq ch(K)$ from K.

The introduction and discussion of the inverse property for f.e. subspaces will be postponed until subsection 2.4. Note that the number of basis functions, or $\dim V(\mathcal{T})$, equals the number of independent interpolation conditions. Quasi-uniformity of $\mathcal{T}$ is not supposed though it would simplify the expression in $(\mathbf{A1})$: the different $\bar{h}_i$ can then be replaced by the one number $h(\mathcal{T})$. The absolute constants occuring in the statements do not depend directly on $\mathcal{T}$ but only on the regularity constant γ_0 (by the way, let us agree on the following convention: c, C will denote absolute constants throughout the exposition (a dependence on p, γ_0, and sometimes also on γ_1 is allowed), for two-sided inequalities such as in $(\mathbf{A1})$ we will often use the notation $\approx$). The somewhat vague property "consisting of locally supported functions" in $(\mathbf{A1})$ may be quantified as follows:

- $(\mathbf{A0})$ For each fixed i, the number of basis functions N_j with support intersecting the support of N_i in a set of positive $\mathbf{R}^d$ measure is bounded by an absolute constant (depending possibly on γ_0).

As a number of investigations show (e.g. [JM, JL, BDR, Li]) it is not necessary to assume locality as strong as in $(\mathbf{A0})$ to get results on the approximation power but this simplifies many proofs (for f.e. subspaces $(\mathbf{A0})$ is obviously fulfilled). In what follows we will not make a difference in this respect: if we speak about $(\mathbf{A1})$ we automatically assume $(\mathbf{A0})$.

The proof of $(\mathbf{A1})$ relies on the affine-linear construction of the f.e. spaces (in more general cases, it may also be true): map any fixed subdomain onto the standard one, the local polynomial space and the interpolation problem as well as the few non-zero basis functions N_i corresponding to this subdomain transform to a fixed interpolation problem on a fixed finite-dimensional polynomial space with a fixed basis on the standard subdomain. The coefficient norm and the L_p-norm are equivalent, with some absolute constants, on the standard subdomain. Transforming back, and adding all local estimates, $(\mathbf{A1})$ comes out. The scaling factors come from the affine-linear mapping, due to the regularity assumption they can be given in terms of the diameters of the supports of the corresponding basis functions (instead of the diameters of triangles itself).

(A2) will be used in 2.3 to derive the approximation order obtainable from f.e. subspaces $V(\mathcal{T})$. It is worth mentioning that the use of quasi-interpolants is more recent: Ciarlet's book, for instance, exclusively uses the natural interpolation projection onto $V(\mathcal{T})$ which is not necessarily well-defined with respect to L_p-spaces (compare, however, [Cl] for the early use of quasi-interpolants in a f.e. context).

To give a typical example, we will construct a quasi-interpolant for linear finite elements. To be precise, consider an arbitrary nodal point P_i (= vertex) of a tetrahedral partition $\mathcal{T}$ of $\Omega \subset \mathbf{R}^d$, and fix a simplex from $\mathcal{T}$ which is at a distance $\leq c\bar{h}_i$ from P_i (with a constant c independent of i !). This simplex will be denoted by Δ_i. See Figure 3 for a possible picture in the two-dimensional case. Due to the regularity requirement, the diameter of Δ_i satisfies $h(\Delta_i) \approx \bar{h}_i$,

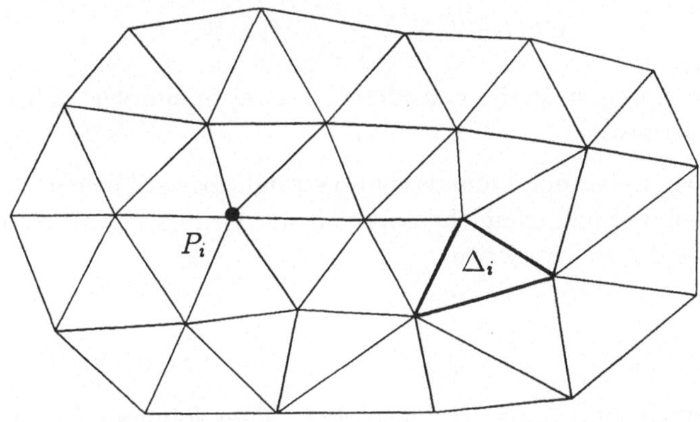

Figure 3. Choice of Δ_i

and its d-dimensional volume $|\Delta_i| \approx \bar{h}_i^d$. Let Q_{ij} resp. $\lambda_{ij}(P)$ denote the vertices of Δ_i resp. the barycentric coordinates of a point $P \in \mathbf{R}^d$ with respect to Δ_i, $j = 0,\ldots,d$. Obviously, the functions λ_{ij} span the space of linear polynomials in $\mathbf{R}^d$, and a straightforward calculation shows that the functions

$$\nu_{ij}(P) = \begin{cases} \frac{d+1}{|\Delta_i|}\left((d+1)\lambda_{ij}(P) - \sum_{k \neq j} \lambda_{ik}(P)\right) &, \quad P \in \Delta_i \\ 0 &, \quad P \notin \Delta_i \end{cases}$$

are L_2-biorthonormal to the basis λ_{ij} of this polynomial space. Finally, we define the functions

$$\nu_i(P) = \sum_{j=0}^{d} \lambda_{ij}(P_i)\nu_{ij}(P) ,$$

with support on Δ_i. Now we are able to write down a formula for a suitable quasi-interpolant:

$$Qf(P) = \sum_i (\int_{\Delta_i} f\nu_i \, dx) N_i(P) .$$

Let us check **(A2)** for this mapping Q. Since the ν_i belong to $L_\infty(\Omega)$, Q is well-defined on $L_p(\Omega)$, $1 \le p \le \infty$, and it maps into V. Due to the above construction and to the assumed regularity of $\mathcal{T}$, there is a constant c such that for any simplex $K \in \mathcal{T}$ the enlarged region $\tilde{K}$ (see **(A2)**) contains all those simplices Δ_i corresponding to the vertices of K. Moreover, $|\lambda_{ji}(P_i)| \le C$, uniformly in i and j. Thus,

$$\begin{aligned}
\|Qf\|_{L_p(K)}^p &\le Ch(K)^d \sum_{P_i \in K} \left| \int_{\Delta_i} f\nu_i \, dx \right|^p \\
&\le Ch(K)^{d(1-p)} \sum_{P_i \in K} (\int_{\Delta_i} |f| \, dx)^p \le C \int_{\tilde{K}} |f|^p \, dx
\end{aligned}$$

for $1 \le p < \infty$, for $p = \infty$ the consideration is even simpler. This proves the local L_p-boundedness.

That Q preserves linear polynomials is also straightforward by construction. Let p be a linear polynomial, using the barycentric coordinates with respect to any of the simplices Δ_i, we can write

$$p(P) = \sum_{j=0}^{d} \alpha_{ij} \lambda_{ij}(P)$$

with some coefficients α_{ij}, and evaluate Qp at P_i as follows:

$$Qp(P_i) = \int_{\Delta_i} p\nu_i \, dx = \sum_{j=0}^{d} \alpha_{ij} \lambda_{ij}(P_i) = p(P_i)$$

(here, the definition of ν_i and the biorthonormality of the systems $\{\lambda_{ij}\}$ and $\{\nu_{ij}\}$ comes in). As a consequence, for any simplex $K \in \mathcal{T}$ we have local preservation of linear polynomials in the following sense: if f coincides with a linear polynomial on $\tilde{K}$ then $Qf = f$ on K. This property implies **(A2)** (by definition of the $\tilde{K}$ and ν_i, it is actually equivalent to **(A2)**).

Choosing appropriate Δ_i, we can adapt the construction of the quasi-interpolant to the particular situation, see subsection 4.2.2 for an example. Usually, some simplex with vertex P_i will do the job. In this case, we even get projections onto $V = V(\mathcal{T})$, i.e., $Q : L_p(\Omega) \to V(\mathcal{T})$ preserves $V(\mathcal{T})$. Also, instead of choosing simplices, we could have taken balls of an appropriate size and distance from P_i as supports for the functions ν_i in the definition of the functionals involved. Though quasi-interpolant constructions have been widely explored in recent years (especially for univariate and multivariate splines, also in the context of shift-invariant subspaces and wavelet theory, see [Bo2, Bo2]), there seems to be no convenient reference for L_p-quasi-interpolants onto general f.e. subspaces over arbitrary partitions which covers the subject generally. No doubt such constructions are possible, one can follow the same procedure as in the above example for the case of linear elements, and reduce the question to a finite-dimensional problem of constructing local biorthonormal systems of appropriate piecewise polynomials for the fundamental system of the interpolation problem on a standard subdomain. We leave this as an exercise, the reader may believe that for all f.e. schemes used in our exposition, property (A2) holds true. In addition, we refer to the traditional construction of natural interpolation projections as in Ciarlet's book [Ci] and their generalizations involving the constructions of dual functionals on the basis of function and derivative evaluations (for approximation processes in spaces of smooth functions, see, e.g., [CL] and the references in [Bo2, Su2]).

Note finally that the properties (A1) and (A2) remain valid uniformly for sequences $\{V_j\}$ of f.e. subspaces whenever the regularity assumption is valid (with γ_0 fixed for the whole sequence). The sequences of f.e. spaces which are the focus of our notes are as a rule produced by first defining an increasing sequence of partitions by some sort of regular dyadic refinement from a rough initial partition. Have in mind that not in all cases does this lead to the monotonicity

$$V_0 \subset V_1 \subset \ldots \subset V_j \subset \ldots$$

of the sequence of subspaces itself, a property which is desirable to avoid a lot of technical, sometimes even principal problems. Examples with this type of defect are provided by the Argyris elements, the Clough-Tocher macroelements, by the serendipity elements and by all non-conforming elements. Last philosophical comment: the reader may have observed that in the formulation of the main properties (A1) and (A2) of f.e. subspaces, we emphasize the role of the basis functions (not of the partitions and local interpolation procedures which led to their construction). This non-traditional approach to finite element theory seems to become more and more popular, it is natural in some respect, and helps to understand finite elements, splines, wavelets, and other species of approximating functions from a unified point of view. It is also important to better understand and investigate multilevel adaptive methods as discussed in sections 4 and 5.

2.1.2 Spline spaces

Univariate piecewise polynomial spaces are quite well-studied, see [Bo1, Su1] for a mathematical introduction. The general definition uses three ingredients: a partition π of an interval $I = [a, b]$ into subintervals $I_i = [x_{i-1}, x_i]$ of length h_i, the vector $\mathbf{m}$ of maximal polynomial order (= degree+1) m_i allowed on each I_i, and the vector $\mathbf{d}$ of smoothness order $d_i = -1, 0, \ldots, \max(m_i, m_{i+1}) - 2$ required at each interior knot x_i. The resulting linear subspace of $L_\infty(I)$ will be denoted by $S_{\mathbf{m}}^{\mathbf{d}}(\pi)$. In most of the practical applications, the case $m_i = m$, $d_i = d$ is needed, in this case we will write $S_m^d(\{I_i\})$. The case $d = -1$ represents non-smooth piecewise polynomial functions of order m on π, for $d = m - 2$ we have maximally smooth splines, the cases $d = 0, 1$ are useful for univariate f.e. applications. There are several standard sets of basis functions for $S_{\mathbf{m}}^{\mathbf{d}}(\pi)$, most important are truncated power functions and B-splines. The latter have minimal local support and possess good numerical properties. For an introduction to B-splines and their properties we refer to [Bo1, Su1].

It took some time to properly generalize the univariate concepts to higher space dimensions. Tensor product constructions on rectangular partitions and regions belong to the prehistory of multivariate spline theory (which does not mean that they have lost their practical attractiveness). Serious progress was made during the eighties (cf. [DM1, Bo1, Ch1]). We characterize the two main directions. One way is to generalize the above definition as follows: Consider, for integer $m = 1, 2, \ldots$, $r = -1, 0, \ldots, m - 2$, and any polyhedral partition $\mathcal{T}$ of Ω,

$$S_m^r(\mathcal{T}) = \{g \in C^r(\Omega) \mid g|_K \in \mathbf{P}_m \ \forall \, K \in \mathcal{T}\}$$

(for $r = -1$ no global smoothness requirement will be made). This definition automatically ensures, for instance, that for any increasing sequence of partitions $\{\mathcal{T}_j\}$ we have

$$\mathbf{P}_m \subset S_m^r(\mathcal{T}_0) \subset S_m^r(\mathcal{T}_1) \subset \ldots \subset S_m^r(\mathcal{T}_j) \subset \ldots \, .$$

This is in contrast to many of the f.e.subspace constructions where incomplete polynomial subspaces resp. non-nested sequences appear very often.

The problem with this approach lies in the difficult questions of determining the dimension and a proper basis, hopefully consisting of locally supported basis functions. Serious problems arise for $r \geq 1$ and small m ($r = 1, m = 3$ is the most investigated case). See [Su2] for some overview. There are many papers which try to get splines and f.e. functions into one theory (look for vertex and super splines, [CL, Su3]). To analyze multivariate spline spaces it is common sense to use Bernstein-Bezier representations of local polynomial patches on simplicial subdomains, instead of writing down the formulae for shape functions in terms of barycentric coordinates (see [Ci]) one has to do with sets of Bezier points and ordinates. These techniques have been developed (both theoretically and algorithmically) mainly for applications of splines for surface representation in CAGD. We refer to [BFK, Fa, HL, Ch2].

The other direction puts basis functions onto the first place. Using geometric and algebraic devices, one defines sets of multivariate B-splines $\{B_i\}$ (piecewise polynomials of order $\leq m'$, automatically possessing a certain global smoothness, with local support) and then investigates the linear subspace they span. The notion of multivariate B-splines goes back to de Boor and then evolved in the early eighties, see [DM1, Bo2, Ch1]. The main problem is to characterize optimal parameters m, m', r such that for a certain partition $\mathcal{T}$

$$\mathbf{P}_m \subset S(\{B_i\}) \equiv \operatorname{span}\{B_i\} \subset S_{m'}^r(\mathcal{T}) \, .$$

Closely related to this problem is the question of local and global linear independence of the B_i in $S(\{B_i\})$, and the existence of polynomial preserving quasi-interpolants (this is necessary to prove properties like **(A1)** and **(A2)**). It turns out, despite intuition, that the approximation order obtainable from the subspace $S(\{B_i\})$ in L_p-norms may be different from the optimal m characterized by the above imbedding, a problem which caused a lot of investigations, see [Bo2, BDR, BJ, BHR].

The dilemma of the second approach is as follows: or the underlying partition $\mathcal{T}$ is very complicated but the subspace $S(\{B_i\})$ relatively rich and easy to handle with (these are the original B-spline constructions of Dahmen/Micchelli [DM1]) or the partition is simple but the structure of $S(\{B_i\})$ is more complicated.

The latter is the case for box splines (see [DM1, Hö1, Hö2, Bo2, Ch1] and the recent book by deBoor/Höllig/Riemenschneider [BHR]) which are conceptionally close to wavelets and use the shift-invariant setting. In the simplest cases, box splines coincide with f.e. nodal basis functions with respect to uniform partitions. To give an example, let us define the family of multivariate box splines associated with a three-directional mesh in the plane $(d = 2)$. Let e_1, e_2 denote the unit coordinate vectors, set $e_3 = e_1 + e_2$ for the vector in diagonal direction, and fix three integers $m_1, m_2, m_3 \geq 1$. Consider the set

$$V \equiv \{v_k\} = \{\underbrace{e_1, \ldots, e_1}_{m_1\ times}, \underbrace{e_2, \ldots, e_2}_{m_2\ times}, \underbrace{e_3, \ldots, e_3}_{m_3\ times}\}$$

of $\tilde{m} = m_1 + m_2 + m_3$ vectors, and define the corresponding box spline geometrically by

$$B_V(x) = \text{vol}_{\tilde{m}-2} \left\{ t \in Q \equiv [0,1]^{\tilde{m}} \mid P_V t \equiv \sum_k t_k v_k = x \right\},$$

as a function of $x \in \mathbf{R}^2$. For other equivalent definitions (using the Hermite-Genocchi formulae, recurrence formulae or Fourier transforms), see [Hö2, BHR]. Finally, $B_{V,k,\mathbf{i}} = B_V(2^k x - \mathbf{i})$, $\mathbf{i} \in \mathbf{Z}^2$, gives the set of all those box-splines on level $k \in \mathbf{Z}$. By these definitions we obviously have

$$\ldots \subset V_{-1} \subset V_0 \subset V_1 \subset \ldots \subset V_k \equiv S(\{B_{V,k,\mathbf{i}}\}) \subset \ldots$$

for the subspaces V_k of linear combinations of box splines of level k which turn out to consist of piecewise polynomial functions of order $\leq m' = \tilde{m} - 1$ with respect to a three-directional partition of $\mathbf{R}^2$ with stepsize 2^{-k}. The support of the box spline B_V is given by $P_V Q$. By induction in m, one can check $B_V \in C^r(\mathbf{R}^2)$, with $r + 2$ the minimal number of vectors in a subset $W \subset V$ such that $V \backslash W$ does not span $\mathbf{R}^2$. The set of box-splines with fixed k forms a locally linear independent and L_p-stable algebraic basis of V_k, i.e. property **(A1)** is satisfied.

These definitions and properties generalize to systems $V \subset \mathbf{R}^d$, with span $\{V\} = \mathbf{R}^d$ and $|V| = n + d$, convex polyhedra $Q \subset \mathbf{R}^{n+d}$ and to even more general situations, see e.g. [Hö2, BHR]. Note that some special cases are quite familiar f.e. spaces with respect to three-directional partitions: $m_1 = m_2 = m_3 = 1$ gives the Courant hat function and linear C^0 finite elements. An interesting box spline leading to a quadratic C^1 element construction on a uniform criss-cross partition is produced by the set $V = \{e_1, e_2, e_1 - e_2, e_1 + e_2\} \subset \mathbf{R}^2$, see [BHR]. We will not go into further details, and refer the interested reader to the above-mentioned references for more information.

2.1.3 Wavelets

Wavelets are an excellent tool for approximating functions (signals, measurements, etc.) on domains with shift-invariance. The adaption to arbitrary domains is under consideration but not extremely promising (both with respect to theory and efficient algorithms). There are several slightly different approaches to wavelet theory. We prefer to explain first what a multiresolution analysis is, and then to clarify the role of the refinement equation in this business, and how to get wavelets and prewavelets out of it. All this can be found in a growing volume of introductory and advanced literature on the subject, see [Db, Ch2, Ch3, ÇDM, Li].

A sequence $\{V_j\}$ of closed subspaces in $L_2(\mathbf{R}^d)$ is called multiresolution analysis (in the sense of Mallat) if the following set of requirements is fulfilled (different authors prefer different but almost equivalent formulations):

- $\ldots \subset V_{-1} \subset V_0 \subset V_1 \subset \ldots \subset V_k \subset \ldots \subset L_2(\mathbf{R}^d)$

- $\bigcap_{k=-\infty}^{\infty} V_k = \emptyset$, $\bigcup_{k=-\infty}^{\infty} V_k$ is dense in $L_2(\mathbf{R}^d)$

- $g(x) \in V_k \iff g(2x) \in V_{k+1}$

- There is a single function $\phi(x) \in V_0$ such that the set of integer translates $\{\phi(x-\mathbf{i}), \mathbf{i} \in \mathbf{Z}^d\}$ forms a Riesz basis in V_0, i.e., $\{\phi(x-\mathbf{i})\}$ is dense in V_0, and

$$\left\| \sum_{\mathbf{i}} c_{\mathbf{i}} \phi(\cdot - \mathbf{i}) \right\|_{L_2(\mathbf{R}^d)} \approx \|\{c_{\mathbf{i}}\}\|_{l_2} \quad \forall \{c_{\mathbf{i}}\} \in l_2 .$$

First examples are provided by multiresolution analyses generated by box-splines. In this case, the proof of the above properties is almost straightforward. The latter property reminds us of $(\mathbf{A1})$ in section 2.1.1 (for $p = 2$). It may be replaced by the stronger requirement that a function $\phi(x) \in V_0$ exists, the integer translates of which form an orthonormal basis of V_0 (in the above mentioned case of box-spline spaces, this new generating function does not have compact support (except the trivial case of $m_1 = m_2 = 1, m_3 = 0$ which leads to piecewise constant functions, the corresponding box-spline is the characteristic function of the unit square)). Note that by dilation the Riesz basis (or orthonormal basis) of V_0 may be transported to any of the V_k, and that the setting is *shift-invariant* in the following sense:

$$g(x) \in V_k \iff g(x - 2^{-k}\mathbf{i}) \in V_k \quad \forall \mathbf{i} \in \mathbf{Z}^d \ \forall k \in \mathbf{Z} .$$

Scales of shift-invariant subspaces generated by one or several functions are subject of recent research activity. Especially, stable bases and L_p-approximation from those subspaces have been studied, see [JM, JL]. Fourier transform techniques are basic for these purposes.

One may use a multiresolution analysis directly for approximation and discretization purposes. However, many efforts have been spent on deriving an orthogonal system with various properties from a multiresolution analysis $\{V_k\}$. This allows us to reduce redundancy in applications and to work with unique decompositions of functions (L_2-orthogonality will play a certain role also in the theory of multilevel methods for variational problems in the Sobolev scale).

More precisely, the strategy is as follows: Consider the orthogonal complement W_1 of V_0 in V_1, construct $\{W_k\}$ by dilation, and observe that

$$L_2(\mathbf{R}^d) = \bigoplus_{k \in \mathbf{Z}} W_k \; .$$

Now, the theory starts to differ for $d = 1$ and $d > 1$. Consider first $d = 1$. To get the desired orthonormal system, one attempts to find a new function $\psi(x) \in W_1 \subset V_1$ the integer translates of which form an orthonormal basis for W_1, and to dilate it to the other levels correspondingly. Any such function ψ is called *wavelet* associated with the multiresolution analysis $\{V_k\}$. If one does not require the orthogonality of the integer translates of ψ but only assumes that they form a Riesz basis in W_1 one arrives at *prewavelets* (or *semiorthogonal wavelets*) associated with the multiresolution analysis. The resulting prewavelet system is automatically a Riesz basis in $L_2(\mathbf{R})$, with partial orthogonality between different levels k. For $d > 1$, one function is not enough: one needs $2^d - 1$ functions the integer shifts of which generate a complete orthonormal system for W_1. The theory for $d = 1$ is in good shape, for $d > 1$ there are still some epsilons left open. We refer to the current literature. Figure 4 b) shows a univariate prewavelet associated with the multiresolution scale of linear splines on uniform dyadic partitions where ϕ is the usual hat function (see Figure 4 a)). The corresponding wavelet may be obtained by orthonormalization of this prewavelet Riesz basis which is equivalent to finding the inverse of a symmetric positive definite, banded bi-infinite Toeplitz matrix. The resulting ψ does not have compact support but still decays exponentially. The corresponding calculations can be carried out more or less explicitly for this simple example. However, in general, it is recommended to deal with the refinement equation and some algebraic consequences of it which lead to efficient recurrence relations governing all computations and also part of the wavelet theory. Let us recall one consequence of the properties of a multiresolution analysis (for simplicity, let $d = 1$): Since $\phi(x) \in V_0 \subset V_1$, and $\{\phi(2x - l)\}$ is the Riesz basis for V_1 we have

$$\phi(x) = \sum_l p_l \phi(2x - l)$$

for some $\{p_l\} \in l_2$. Applying the Fourier transform to both sides one gets a reformulation of this refinement equation as

$$\hat{\phi}(\xi) = P(e^{-i\xi/2})\hat{\phi}(\xi/2) \, , \qquad P(z) = \frac{1}{2} \sum_l p_l z^l \, ,$$

with $P(z)$ called symbol of ϕ. For instance, for the linear B-spline shown in

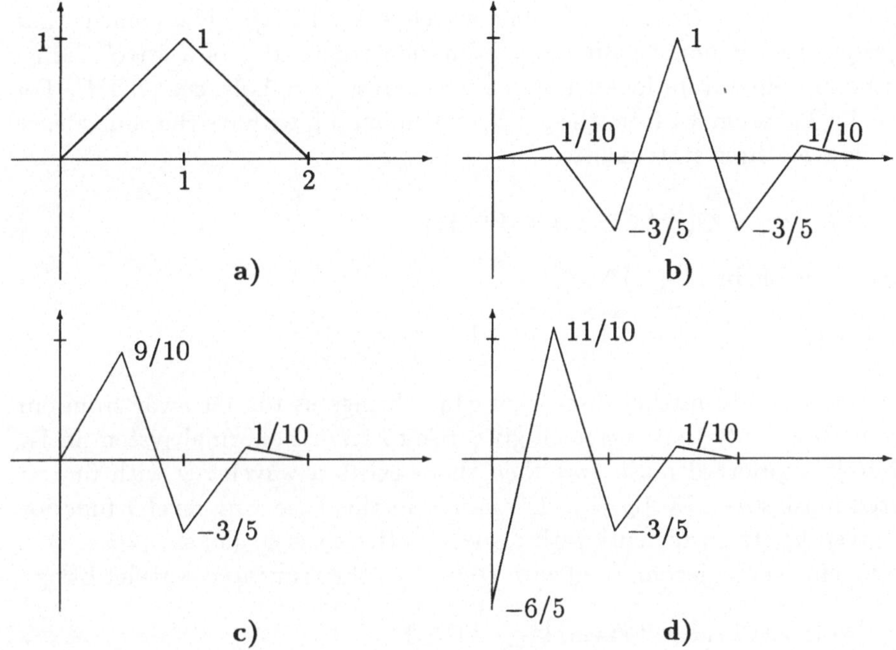

Figure 4. Linear B-spline and prewavelets

Figure 4 a) the symbol is a polynomial

$$P(z) = \frac{1}{4}(1+z)^2 \quad (p_0 = p_2 = 1/2,\, p_1 = 1,\, p_l = 0 \text{ otherwise}) .$$

Analogous formulae should hold for prewavelets and wavelets:

$$\psi(x) = \sum_l q_l \phi(2x - l)$$

for some $\{q_l\} \in l_2$, or, equivalently,

$$\hat{\psi}(\xi) = Q(e^{-i\xi/2})\hat{\phi}(\xi/2) , \qquad Q(z) = \frac{1}{2}\sum_l q_l z^l ,$$

with $Q(z)$ the symbol of ψ. E.g., for the prewavelet shown in Figure 4 b) we have

$$Q(z) = \frac{1}{20}(z^4 - 6z^3 + 10z^2 - 6z + 1) = \frac{1}{20}(z-1)^2(z^2 - 4z + 1)$$

(for shift-invariant spline spaces , there is a general prewavelet theory using finitely supported masks $\{p_l\}$, $\{q_l\}$, a property which is numerically desirable but not fulfilled for fully orthonormal spline wavelets, see [Ch2]). The point is that many properties (orthonormality, approximation order, etc.) of a wavelet scheme can be obtained from looking at the masks resp. symbols, see [CDM]. For instance, by Theorem 3.1 from the paper by Chui in [Li] we have the equivalence of the following three statements:

- $\inf_{g \in V_k} \|f - g\|_2 = O(2^{-km}) \quad \forall f \in W_2^m(\mathbf{R})$

- $P(z)$ is divisible by $(1 + z)^m$

- $\sum_l (-1)^l l^r p_l = 0 \quad , \quad r = 0, 1, \ldots, m - 1.$

We will not step into further discussion which brings us too far away from our main goal. Note only that if ϕ itself gives rise to an orthonormal system in V_0, with finitely supported mask $\{p_l\}$ then there exists a wavelet ψ with finitely supported mask given by $\{q_l = (-1)^l \cdot p_{1-l}\}$. In this case, any level j function $g \in V_j$, given by its coefficients with respect to the basis $\phi_{j,i}(x) \equiv \phi(2^j x - i)$ of V_j, can be efficiently decomposed with respect to the truncated wavelet basis

$$\{\phi_{k,i}\} \cup \{\psi_{k+1,i}\} \cup \{\psi_{k+2,i}\} \cup \ldots \cup \{\psi_{j,i}\}$$

of the same $V_j = V_k \oplus W_{k+1} \oplus W_{k+2} \oplus \ldots \oplus W_j$ vice versa (the part of this decomposition corresponding to V_k may be neglected if k is a sufficiently large negative integer, cf. the properties of the multiresolution analysis). These decomposition problems are the basis to any kind of applied Fourier analysis and synthesis, and their efficient solution has important consequences in signal analysis etc.. The existence of such ϕ was first shown by Daubechies, see [Db, Ch2]. If one does not insist on having a completely orthonormalized system, things will be much easier: prewavelets (or semi-orthogonal wavelets in the terminology of [Ch2]), biorthogonal wavelets and frames allow for more flexibility, and lead to masks with small finite support. Below we will deal with the approximate solution of boundary value problems for p.d.e's requiring the introduction of essential boundary conditions into the definition of the subspaces. This task is by no means simple for orthonormal wavelets; a bit easier is the construction of pre-wavelets satisfying boundary conditions. An example, once again for univariate linear splines, is given in Figure 4 c) resp. d). It illustrates the boundary modification of the linear prewavelet of Figure 4 b) if homogeneous Dirichlet resp. Neumann boundary conditions are imposed. We will come back to the problem of wavelets for p.d.e. solving later on.

2.2 Moduli of smoothness and K-functionals

We are dealing with measuring L_p-smoothness. Classical results use Lipschitz and Hölder continuity, conditions that explore first order differences and fill the gap between C^0 and C^1. A natural generalization to L_p and arbitrary smoothness are the L_p moduli of smoothness. Throughout this and the following subsections, we suppose $1 \leq p \leq \infty$, for $p = \infty$, however, we replace $L_\infty(\Omega), W_\infty^m(\Omega)$ without further mentioning by $C(\bar\Omega), C^m(\bar\Omega)$.

Let $\Omega \subset \mathbf{R}^d$ be open, for arbitrary $m = 1, 2, \ldots$ and $h \in \mathbf{R}^d$ define the subset

$$\Omega_{m,h} = \{x \in \mathbf{R}^d \mid [x, x + mh] \subset \Omega\}.$$

Let

$$I_m(h, f)_p = \|\Delta_h^m f\|_{L_p(\Omega_{m,h})}$$

be the L_p-norm of the directional m-th difference

$$\Delta_h^m f(x) = \sum_{r=0}^m (-1)^{m-r} \binom{m}{r} f(x + rh)$$

of $f \in L_p(\Omega)$. We define the total m-th order L_p-modulus of smoothness of f by

$$\omega_m(t, f)_p = \sup_{|h| \leq t} I_m(h, f)_p \ , \qquad t > 0 \ ,$$

and the corresponding directional moduli with respect to the direction $e \in \mathbf{R}^d$ (of unit length) by

$$\omega_m^e(t, f)_p = \sup_{0 < \tau \leq t} I_m(\tau e, f)_p \ , \qquad t > 0 \ .$$

For tensor-product domains, the so-called partial moduli of smoothness $\omega_m^{e_i}(t, f)_p$ with respect to the i-th coordinate direction e_i, $i = 1, \ldots, d$, are of special interest. We will not concentrate on more complicated moduli $\omega_{m_1,\ldots,m_d}^{e_1,\ldots,e_d}(t, f)_p$ which use mixed differences $\Delta_{h_1 e_1}^{m_1} \ldots \Delta_{h_d e_d}^{m_d} f(x)$ in their definition.

Using properties of differences, one can prove many properties of these moduli in an elementary way, at least for domains with nice properties. The most elegant way is to first prove everything for a standard domain such as $\mathbf{R}^d$, and then to carry over results to domains for which extension theorems hold true, especially such extension procedures which preserve the order of the corresponding L_p-moduli (the extension trick is quite convenient in the theory of function spaces, see 3.2). Some of such properties are valid for any domain, and there is no need to use extension theorems. The properties in the list below are of this type:

- $\omega_m(t, f)_p \to 0$ for $t \to 0$ $\forall f \in L_p(\Omega)$.

- $\omega_m(t, f)_p$ is nondecreasing and positive (for $f \notin \mathbf{P}_m$) on $t > 0$.

- $t^{-m}\omega_m(t, f)_p$ is almost decreasing, i.e. there exists a constant c such that

$$\tau^{-m}\omega_m(\tau, f)_p \geq ct^{-m}\omega_m(t, f)_p , \quad 0 < \tau < t .$$

Thus, m-th order moduli of smoothness have saturation order m :

$$\omega_m(t, f)_p = o(t^m) , \ t \to 0 \quad \Rightarrow f \in \mathbf{P}_m$$

- $\omega_m(t, f + g)_p \leq \omega_m(t, f)_p + \omega_m(t, g)_p$

- $\omega_m(t, f)_p \leq 2\omega_{m-1}(t, f)_p \leq \ldots \leq 2^m\|f\|_p$

- We have

$$\omega_m^e(t, f)_p \leq Ct\omega_{m-1}^e(t, \frac{\partial f}{\partial e})_p , \ t > 0$$

for any function with directional derivative $\frac{\partial f}{\partial e} \in L_p(\Omega)$, as a consequence

$$\omega_m(t, f)_p \leq Ct^m|f|_{m,p} \quad \forall f \in W_p^m(\Omega) .$$

For proofs and more properties (some of them will be mentioned below) in the univariate case ($\Omega = [a, b]$), see [Ti, LD], the multivariate case can be found in [Ni, BIN, JS2, DDS].

In the latter two references, moduli of smoothness are compared to the so-called K-functional between L_p and W_p^m arising from interpolation theory, see also [BL, Tr2] for some generalities on interpolation methods in Banach and function spaces, and [LD] for an exposition in the univariate case . We give the definition:

$$K_m(t, f)_p = \inf_{g \in W_p^m(\Omega)} \{\|f - g\|_p + t^m|g|_{m,p}\} , \quad t \to 0$$

is called m-th order K-functional for $f \in L_p(\Omega)$. In [JS2], the following theorem is contained (for modifications involving combinations of partial and mixed moduli of smoothness, see [DDS]) :

Theorem 1. (Equivalence of moduli of smoothness and K-functionals) Let Ω satisfy the uniform cone condition. Then, for any $m = 1, 2, \ldots, 1 \le p < \infty$, and $f \in L_p(\Omega)$, we have

$$\omega_m(t, f)_p \approx K_m(t, f)_p \quad , \qquad t > 0 \, .$$

The assertion holds for $p = \infty$, with C and C^m replacing L_p and W_p^m, resp..

The uniform cone condition is defined, e.g., in [Ad] (together with other geometrical conditions which arise in connection with classes of differentiable functions on domains, cf. also [Gv, Nč, Sp, TW]). For bounded domains it consists of the existence of a finite open covering $V_1, \ldots V_N$ of $\partial\Omega$ and a collection of fixed finite cones $L_1, \ldots, L_N$ such that $x \in U_i \cap \Omega$ implies $x + L_i \subset \Omega$ where $U_i = \cup_{x \in V_i} B(x, \delta)$ denotes the open δ neighbourhood of V_i for some $\delta > 0$ ($B(x, r)$ is the d-dimensional open ball with center at x and radius $r > 0$). Obviously, almost all polyhedral domains satisfy this condition, the exception are slit domains where the problem occurs with the tip of the slit.

Remark. Theorem 1 is a technically very helpful equivalence which, roughly speaking, allows us to reduce estimates involving moduli of smoothness (yielding results on the fine smoothness scale) to simpler norm estimates for smooth functions from W_p^m (only integer smoothness parameter). In contrast to the historical evolution of this part of approximation theory where first moduli of smoothness were basic, nowadays the K-functional approach is much more convenient and often preferred (see [Del, LD] for some one-dimensional examples in this direction). Nevertheless, the moduli are elementary, and some results are easier to get by using them. This will be illustrated in the next subsections.

2.3   Jackson and Whitney inequalities

Given a subset $M \in L_p(\Omega)$, consider the *best approximation*

$$E_M(f)_p = \inf_{g \in M} \|f - g\|_p \quad , \qquad f \in L_p(\Omega).$$

As a rule, we are dealing with linear subspaces rather than with general sets M. The study of these best approximations (both with respect to considering the above given extremal problem for individual functions as well as for classes of functions, and their asymptotic properties for sequences of subsets M) is the heart of classical approximation theory.

We are interested in getting estimates from above for $E_M(f)_p$ when M coincides with a f.e. subspace of subsection 2.1.1.

Theorem 2. (Jackson-type inequality) Let $\mathcal{T}$ be any regular partition of a polyhedral domain $\Omega \subset \mathbf{R}^d$ satisfying the uniform cone condition, and let $V(\mathcal{T})$ be an f.e. subspace as defined in section 2.1.1 satisfying property **(A2)**. Then we have

$$\textbf{(J)} \qquad E_{V(\mathcal{T})}(f)_p \leq \|f - \mathcal{Q}f\|_p \leq C\omega_m(h(\mathcal{T}), f)_p \qquad \forall f \in L_p(\Omega)$$

where m is defined in **(A2)**, and the absolute constant C depends on the constants in **(A2)**.

We sketch two of three known (and slightly different) methods of proof of this type of assertion. According to **(A2)**, for each subdomain K and any polynomial $p_m \in \mathbf{P}_m$ we have

$$\|f - \mathcal{Q}f\|_{L_p(K)} \ \leq \ \|f - p_m\|_{L_p(K)} + \|\mathcal{Q}(f - p_m)\|_{L_p(K)}$$

$$\leq \ (1 + C)\|f - p_m\|_{L_p(\tilde{K})}$$

which reduces the estimates to the problem of local best approximation by algebraic polynomials. Its solution in terms of local moduli of smoothness is one of the basic estimates of L_p approximation theory, and goes back to Whitney for univariate functions. Information about the multivariate case is given in [JS2, DDS, BI]. We quote a corollary :

Theorem 3. (Whitney-type inequality) Let $\Omega \subset \mathbf{R}^d$ satisfy the uniform cone condition. Then for all $m = 1, 2, \ldots,$ $1 \leq p \leq \infty$, and $f \in L_p(\Omega)$

$$\textbf{(W)} \qquad E_{\mathbf{P}_m}(f)_p = \inf_{p_m \in \mathbf{P}_m} \|f - p_m\|_p \leq C_{m,\Omega}\omega_m(\mathrm{diam}(\Omega), f)_p$$

The dependence of the constant on Ω comes from the cone condition.

This theorem will now be applied locally (to each $\tilde{K}$). This gives

$$\|f - \mathcal{Q}f\|_p^p \ = \ \sum_K \|f - \mathcal{Q}f\|_{L_p(K)}^p$$

$$\leq \ ((1 + C)C^*)^p \sum_K (\omega_m(\mathrm{diam}(\tilde{K}), f)_{L_p(\tilde{K})})^p$$

C^* is the maximum of the constants $C_{m,\tilde{K}}$, and depends on $C_{m,\Omega}$ and possibly on the regularity constant γ_0. There remains to come to the modulus with respect to Ω. This is tricky, one may act using the equivalence

$$\omega_m(t, f)_p^p \approx t^{-m} \int_{|h| \leq t} I_m(h, f)_p^p, \quad t > 0 \,,$$

which holds under some restriction on Ω but is fulfilled for our $\tilde{K}$. Since different $\tilde{K}$ have finite overlap (see **(A2)** and **(A0)**), the resulting sum can be finally estimated by $c\omega_m(h(\mathcal{T}), f)_p^p$. Though elementary, all these considerations cause a lot of technical difficulties.

We will therefore present a second proof relying on Theorem 1. Let $g \in W_p^m(\Omega)$ be arbitrary. Then

$$\|f - \mathcal{Q}f\|_p \le \|f - g\|_p + \|g - \mathcal{Q}g\|_p + \|\mathcal{Q}(f - g)\|_p \ .$$

Since the local L_p boundedness of the quasi-interpolant assumed in $(\mathbf{A2})$ immediately yields the global $L_p(\Omega)$ boundedness, the third term can be bounded by a constant multiple of the first one. Thus, what remains is to prove an estimate for the second term, i.e. for the approximation order by the quasi-interpolant in the case of smooth functions g (instead of attacking the problem for arbitrary non-smooth f). Comparing with Theorem 1, we see what would be sufficient:

$$\|g - \mathcal{Q}g\|_p \le Ch(\mathcal{T})^m |g|_{m,p} \quad \forall\, g \in W_p^m(\Omega) \ .$$

This can be done as above, using locally $(\mathbf{A2})$ and summing up but avoiding the deep Whitney-type result. Instead, some variant of the Bramble-Hilbert lemma suffices. To be precise, as above one gets for arbitrary $p_m \in \mathbf{P}_m$

$$\|g - \mathcal{Q}g\|_{L_p(K)} \le C \inf_{p_m \in \mathbf{P}_m} \|g - p_m\|_{L_p(\tilde{K})} \le C\mathrm{diam}(\tilde{K})^m |g|_{W_p^m(\tilde{K})} \ .$$

The latter inequality is actually a consequence of the Bramble-Hilbert argument [BH], it can be proved first for smooth functions using the Taylor polynom resp. remainder formulae, and then extended to all of W_p^m by continuity (the density of smooth functions in $W_p^m(\Omega)$ is true for this type of domain, see [Ad, Gv]), more advanced techniques are reported on in [Hu]. Now, we take the sum of the p-th powers of both sides with respect to $K \in \mathcal{T}$, and observe that each K is covered by only a finite number (depending on the regularity constant of $\mathcal{T}$) of $\tilde{K}$'s. This gives the desired estimate.

There is a third, rather elementary, but tedious way to prove Jackson-type estimates for f.e. functions: One relies on the Whitney-type estimate $(\mathbf{W})$ for special domains, say simplices, which follows from the case of a cube [Bd] in an uncomplicated manner, and fits the pieces together by choosing nodal values as averages of the corresponding values from different pieces such that the resulting f.e. function remains well-approximating. The advantage is that one does not need the deep and nontrivial extension results, also the uniform cone condition appears to be superflous (cf. the remark below). See [Os1] where an extension to L_p, $p < 1$, has been obtained along these lines.

Remark. It is possible to extend the estimate of Theorem 2 (but not the result of Theorem 3 !) to more general polyhedral domains which do not satisfy the cone condition. This observation is important for domains with slits occuring sometimes in practice. The argument is simply to consider decompositions of Ω into overlapping domains satisfying the uniform cone condition and such that they are consistent with $\mathcal{T}$. We give an outline for the case of linear triangular C^0 elements ($m = 2$) for a standard slit domain in the plane.

The basic idea generalizes to other situations without problems. Figure 5 contains a fictive triangulation $\mathcal{T}$ of a slit domain Ω into 7 triangles. We form

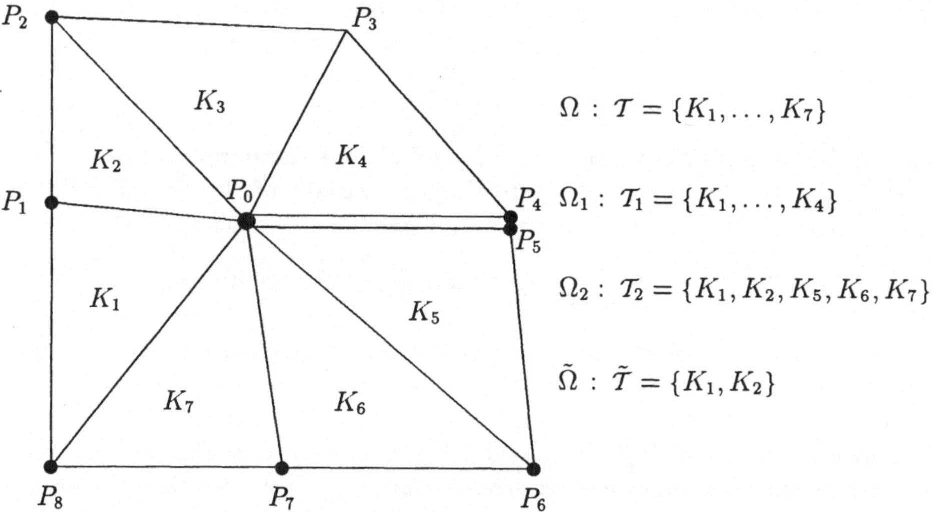

Figure 5. Decomposition of a slit domain

two subtriangulations $\mathcal{T}_1$ and $\mathcal{T}_2$ with common part $\tilde{\mathcal{T}} \neq \emptyset$, the corresponding domains Ω_1 and Ω_2 satisfy now the uniform cone condition. Thus, we can find two approximating linear f.e. functions $g_l \in V(\mathcal{T}_l)$ such that

$$\|f - g_l\|_{L_p(\Omega_l)} \leq C\omega_2(h(\mathcal{T}_l), f)_{L_p(\Omega_l)} \leq C\omega_2(h(\mathcal{T}), f)_p$$

for $l = 1, 2$. It remains to mix these two functions to obtain a good approximant $g \in V(\mathcal{T})$. E.g., we can prescribe the nodal values of g as follows:

$$g(P_i) = \begin{cases} g_l(P_i) & \text{if} \quad P_i \in \bar{\Omega}_l \backslash \bar{\tilde{\Omega}}, \, l = 1, 2 \\ (g_1(P_i) + g_2(P_i))/2 & \text{if} \quad P_i \in \bar{\tilde{\Omega}} \end{cases}$$

Then it immediately follows (cf. **(A1)**) that

$$\|f - g\|_p \;\le\; \sum_{l=1}^{2} \|f - g\|_{L_p(\Omega_l)}$$

$$\le\; \sum_{l=1}^{2}(\|f - g_l\|_{L_p(\Omega_l)} + \|g - g_l\|_{L_p(\Omega_l)})$$

$$\le\; C(\omega_2(h(\mathcal{T}), f)_p + \|g_1 - g_2\|_{L_p(\tilde{\Omega})})$$

$$\le\; C(\omega_2(h(\mathcal{T}), f)_p + \sum_{l=1}^{2} \|f - g_l\|_{L_p(\Omega_l)})$$

$$\le\; C\omega_2(h(\mathcal{T}), f)_p\,.$$

Thus, the best approximation is estimated by the corresponding value of the modulus of smoothness! This in turn gives the estimate for the quasi-interpolant (recall that our construction in 2.1.1 of $\mathcal{Q}$ resulted even in a $L_p(\Omega)$ bounded *projection* onto $V(\mathcal{T})$ which implies

$$\|f - \mathcal{Q}f\|_p \;\le\; \inf_{g \in V(\mathcal{T})}(\|f - g\|_p + \|\mathcal{Q}(f - g)\|_p)$$

$$\le\; (1 + \|\mathcal{Q}\|_{L_p \to L_p})E_{V(\mathcal{T})}(f)_p\ )\,.$$

Note finally, that due to the localization of all estimates the assertion holds for regular partitions of the whole $\mathbf{R}^d$ as well. Quasi-uniformity is not required in the direct estimates, the largest subdomain K determines the rate of approximation (which can easily be shown to be asymptotically correct on the whole class $L_p(\Omega)$).

2.4   Bernstein inequalities and inverse estimates

Bernstein-type inequalities describe the difference resp. differentiability properties of functions from subspaces used in approximation processes. Here we have in mind the behaviour of the moduli of smoothness of finite element functions. As an application, we get inverse estimates for finite element approximation, i.e. we find sharp estimates for the moduli of smoothness of functions $f \in L_p(\Omega)$ in terms of their best approximations with respect to increasing sequences of f.e. subspaces (for another type of such inverse theorems using non-increasing sequences corresponding to triangulations satisfying certain mixing conditions, see [DDS]).

Theorem 4. (Bernstein-type inequality). Let $V(\mathcal{T}) \subset S^r_{m'}(\mathcal{T})$ be a f.e. subspace satisfying property **(A1)**. Then, for any $\bar{m} = 1, 2, \ldots$ there is a constant

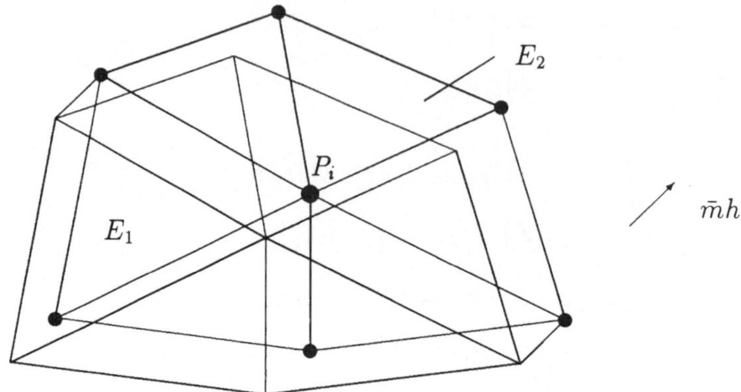

Figure 6. The sets E_1 and E_2

C such that

(B) $\omega_{\bar{m}}(t,g)_p \leq C(\min(1\,,\,t/h_{\min}(\mathcal{T}))^{\min(r+1+1/p,\bar{m})})\,\|g\|_p$

for all $g \in V(\mathcal{T})$ and $t > 0$.

Proof. For $t \geq h_{\min}(\mathcal{T})$ the inequality holds with $C = 2^{\bar{m}}$ (see the properties of the moduli of smoothness quoted in 2.2). Thus, let $h \in \mathbf{R}^d$ be such that $|h| \leq t < h_{\min}(\mathcal{T})$. By the local support property of the nodal basis $\{N_i\}$, we have

$$I_{\bar{m}}(h,g)_p^p \leq C \sum_i |c_i|^p I_{\bar{m}}(h,N_i)_p^p$$

(with C depending on γ_0). Now, fix an arbitrary i. Figure 6 shows the split of $\operatorname{supp}(\Delta_h^{\bar{m}} N_i)$ into two parts E_1 and E_2 (with E_1 being the set of all interior points $x \in \operatorname{supp} N_i$ such that the segment $[x, x+\bar{m}h]$ belongs to only one of the subdomains K of $\mathcal{T}$). Observe that due to the regularity assumption

$$\operatorname{mes}(E_1) \leq C\bar{h}_i^d\,, \qquad \operatorname{mes}(E_2) \leq C\bar{h}_i^{d-1}|h|\quad.$$

On the intersection of E_1 with any fixed subdomain $K \subset \operatorname{supp} N_i$ we can use the property that $\Delta_h^{\bar{m}} N_i$ coincides with the difference of some polynomial $p_m \in \mathbf{P}_m$ with $\|p_m\|_{C(K)} \leq C$, cf. **(A1)** for $p = \infty$. By the Taylor formulae we get

$$\int_{E_1} |\Delta_h^{\bar{m}} N_i(x)|^p\,dx \;=\; \sum_{K \subset \operatorname{supp} N_i} \int_{K \cap E_1} |\Delta_h^{\bar{m}} p_m(x)|^p\,dx$$
$$\leq\; C(|h|/\bar{h}_i)^{\bar{m}p}\operatorname{mes}(E_1)\,.$$

On the other hand, on E_2 we have the $\bar{m}$-th order difference of a piecewise polynomial of global smoothness class C^r (which therefore belongs to W_∞^{r+1} but as a rule not to C^{r+1}). If $\bar{m} \leq r$ then we can estimate as above. For $\bar{m} > r$, we express the $\bar{m}$-th order difference by $(r+1)$-th order ones, use that $|N_i|_{r+1,\infty} \leq C\bar{h}_i^{-(r+1)}$, and finally arrive at

$$\int_{E_2} |\Delta_h^{\bar{m}} N_i(x)|^p \, dx \leq C(|h|/\bar{h}_i)^{(r+1)p} \mathrm{mes}\,(E_1) \ .$$

Now, substituting the estimates for the measure of E_1 and E_2, we have

$$I_{\bar{m}}(h, N_i)_p = \int_{\Omega_{\bar{m},h}} |\Delta_h^{\bar{m}} N_i(x)|^p \, dx \leq C(|h|/\bar{h}_i)^{\min((r+1)p+1,\bar{m}p)} \bar{h}_i^d \ .$$

This yields the assertion of the theorem (the case $p = \infty$, with L_∞ substituted by C, can be handled in analogy).

Now we come to the formulation of the inverse estimate.

Theorem 5. Let $\mathcal{T}_j$ be a sequence of quasi-uniform partitions so that

$$V_0 \subset V_1 \subset \ldots \subset V_j \equiv V(\mathcal{T}_j) \subset \ldots, \quad h(\mathcal{T}_j) \approx a^{-j} \quad (a > 1).$$

Besides that, let the assumptions of Theorem 4 be satisfied. Then we have for all $\bar{m} = 1, 2, \ldots, f \in L_p(\Omega)$, and $j \geq 0$

$$\textbf{(I)} \qquad \omega_{\bar{m}}(a^{-j}, f)_p \leq Ca^{-j\lambda}\left(\|f\|_p + \sum_{l=0}^{j} a^{l\lambda} E_{V_j}(f)_p \right)$$

where $\lambda = \min(\bar{m}, r + 1 + 1/p)$.

Proof. This is an immediate consequence of Theorem 4 and the monotonicity of $\{V_j\}$. By definition of the best approximations there exist f.e. functions $g_j \in V_j$ such that $\|f - g_j\|_p = E_{V_j}(f)_p$, $j \geq 0$. Consider the finite representation

$$f = (f - g_j) + \sum_{l=0}^{j} g_l^*$$

with $g_l^* = g_l - g_{l-1} \in V_l$, $l \geq 1$ due to the monotonicity of $\{V_j\}$ (for $l = 0$ we define $g_0^* = g_0$). Now we apply first some inequalities for moduli of smoothness from 2.2 and then **(B)** (note that there $h_{\min}(\mathcal{T}_j)$ can be changed into $h(\mathcal{T}_j) \approx a^{-j}$ due to the quasi-uniformity of the partitions):

$$\omega_{\bar{m}}(a^{-j}, f)_p \ \leq \ \omega_{\bar{m}}(a^{-j}, f - g_j)_p + \sum_{l=0}^{j} \omega_{\bar{m}}(a^{-j}, g_l^*)_p$$

$$\leq C\left\{E_{V_j}(f)_p + \sum_{l=0}^{j}(a^{-j}/h(\mathcal{T}_l))^\lambda \|g_l^*\|_p\right\}$$

$$\leq C\left\{E_{V_j}(f)_p + \sum_{l=0}^{j}a^{(l-j)\lambda}(E_{V_l}(f)_p + E_{V_{l-1}}(f)_p)\right\}$$

$$\leq C\left\{\|f\|_p + a^{-j\lambda}\sum_{l=0}^{j}a^{l\lambda}E_{V_l}(f)_p\right\}\ .$$

with $E_{V_{-1}}(f)_p \equiv \|f\|_p$.

Remark. Theorem 5 enables us to give upper bounds for difference and differentiability properties of L_p functions from knowledge on the best approximations with respect to certain increasing sequences of f.e. subspaces while Theorem 2 (sometimes called direct theorem of f.e. approximation theory) serves the opposite direction. This is the basis for the characterization of classes of differentiable functions by best approximations with respect to a sequence $V_j \equiv V(\mathcal{T}_j)$ satisfying the assumptions of Theorems 2 and 5, see subsection 2.6 below. Note that for **(I)** we have used the quasi-uniformity of the underlying partitions which was not necessary for the previous theorems. Under mild conditions one can also prove that the λ occuring in Theorem 5 (and implicitely in Theorem 4) cannot be improved.

2.5   Information on other approximation schemes

This is a very short section stating that many spline and wavelet schemes also allow for direct and inverse theorems in terms of moduli of smoothness or K-functionals. For $S_{m'}^r(\mathcal{T})$ possessing a local B-spline-like basis this can be proved in completely the same way as for finite elements. There were results in the 70's (cf. [Sr1, Cs] or the monograph [LD]) on one-dimensional spline approximation as well as on multivariate counterparts (cf. [DDS]).

The main bulk of papers on approximation by shift-invariant subspaces (box-splines, multiresolution analysis/prewavelets/wavelets) concentrates on the direct estimates stating them for functions from Sobolev classes (see especially the papers on controlled L_p-approximation generalizing a Strang-Fix result from 1973, cf. [DM2, Bo2, CDM] for some overviews and more precise information). A moment's reflection shows that this suffices (as in the second proof of Theorem 2) to get estimates involving moduli of smoothness of the corresponding order. The inverse property was recently discussed in [DK], a paper which is close in spirit to the intention of our notes.

Unfortunately, an easily readable reference covering all these topics of approximation theory by local schemes is still not available. Much more elaborated (and more traditional) is the approximation method with respect to trigonometric polynomials on the d-dimensional torus, and to entire analytic functions of exponential type (= Fourier transform supported in a ball) on $\mathbf{R}^d$, see e.g. [Ni, BL, Tr2]. This is (especially the case $d = 1$) the field where the basic techniques and the names of the main theorems (Jackson, Bernstein, etc.) come from, see [Ti, Te, LD] for more information. However, there is a striking difference: the approximation schemes considered in our notes are local in nature which is important for the applications we have in mind.

2.6 Constructive characterization of Besov spaces

We give one of the classical definitions of the Besov spaces of smoothness $s > 0$ on a domain Ω : Let $B_{p,q}^{s,m}(\Omega)$ be the set of all $f \in L_p(\Omega)$ such that the semi-norm

$$|f|_{B_{p,q}^{s,m}} = \begin{cases} \left(\int_0^\infty \omega_m(t,f)_p^q t^{-sq-1}\, dt\right)^{1/q} & \text{for } 1 \le q < \infty \\[2mm] \sup_{t>0} t^{-s}\omega_m(t,f)_p & \text{for } q = \infty \end{cases}$$

is finite. Here $m > s$ is an arbitrarily fixed integer. A norm is defined by

$$\|f\|_{B_{p,q}^{s,m}} \equiv \|f\|_p + |f|_{B_{p,q}^{s,m}}\ \ .$$

Note that for $p = \infty$ it is once again reasonable to replace $L_\infty(\Omega)$ by $C(\bar{\Omega})$ in the definitions. In analogy to the L_p case, one proves that the Besov classes $B_{p,q}^{s,m}(\Omega)$ form a scale of Banach spaces (separability can be proved iff $q < \infty$).

It turns out that under the uniform cone condition on the domain, the spaces are not different for different $m > s$, this follows from the L_p variant of Marchaud's inequality proved in [JS2]:

$$\omega_{m'}(t,f)_p \le Ct^{m'}\left(\|f\|_p + \int_t^\infty \omega_m(\tau,f)_p t^{-m'-1}\, dt\right)\,,$$

for $m' < m$ and $t > 0$. Also, due to the trivial upper estimate of $\omega_m(t,f)_p$ by the L_p-norm of f stated in 2.2, the range $0 < t < \infty$ in the definition can be replaced by any finite interval $0 < t \le t_0$, yielding the same space with an equivalent norm. The next theorem is the output of Theorems 2 and 4/5, the uniform cone condition may be shown to be superfluous, see the remark after Theorem 2.

Theorem 6. (Characterization of Besov classes by f.e. approximation) Let Ω be a polyhedral domain with the uniform cone condition. Suppose that the sequence of f.e. subspaces $\{V_j\} \equiv \{V(\mathcal{T}_j)\}$ satisfies the assumptions of Theorems 2 and 4, i.e. **(A1)**, **(A2)** are fulfilled, the underlying partitions $\mathcal{T}_j$ are quasi-uniform with characteristic stepsize $h(\mathcal{T}_j) \approx a^{-j}$ decaying like a geometric progression, and the subspaces $V_j \subset S_{m'}^r(\Omega)$ form an increasing sequence. Then, for $0 < s < \lambda = \min(m, r + 1 + 1/p)$ the following expressions define equivalent norms on $B_{p,q}^{s,m}(\Omega)$:

$$\|f\|_{B_{p,q}^{s,m}}^+ = \|f\|_p + \|\{a^{js} E_{V_j}(f)_p\}\|_{l_q}$$

$$\|f\|_{B_{p,q}^{s,m}}^* = \inf_{g_j \in V_j \, : \, f = \sum_j g_j} \|\{a^{js}\|g_j\|_p\}\|_{l_q}$$

$$\|f\|_{B_{p,q}^{s,m}}^{**} = \inf_{\{c_{j,i}\} \, : \, f = \sum_j \sum_i c_{j,i} N_{j,i}} \|\{a^{j(s-d/p)}\|\{c_{j,i}\}\|_{l_p}\}\|_{l_q}$$

$$\|f\|_{B_{p,q}^{s,m}}^{***} = \|\{a^{js}\|(R_j - R_{j-1})f\|_p\}\|_{l_q}$$

where $R_j : L_p(\Omega) \to V_j$ is a sequence of uniformly L_p-bounded linear projections onto V_j, $j \geq 0$ ($R_{-1} = 0$). The constants in the norm equivalencies depend on the constants characterizing the regularity and quasi-uniformity of the sequence $\{\mathcal{T}_j\}$, the finite element type used, on a, and on the smoothness parameter s.

Proof. This is a simple exercise. The relations

$$\|f\|_{B_{p,q}^{s,m}}^{**} \approx \|f\|_{B_{p,q}^{s,m}}^* \leq C\|f\|_{B_{p,q}^{s,m}}^{***}$$

are obvious (the first equivalence follows from **(A1)**). Since

$$\|f - R_j f\|_p \leq \|f - g_j\|_p + \|R_j(f - g_j)\|_p \leq C\|f - g_j\|_p$$

for any $g_j \in V_j$, we obtain

$$\|(R_j - R_{j-1})f\|_p \leq \|f - R_j f\|_p + \|f - R_{j-1}f\|_p$$

$$\leq C(E_{V_j}(f)_p + E_{V_{j-1}}(f)_p), \quad j \geq 0$$

(for $j = 0$ one substitutes $E_{V_{-1}}(f)_p = \|f\|_p$). This yields

$$\|f\|_{B_{p,q}^{s,m}}^{***} \leq C\|f\|_{B_{p,q}^{s,m}}^+ .$$

Note that the quasi-interpolant construction given in 2.1.1 provides examples of sequences of projections satisfying the above properties.

Moreover,

$$\|f\|^{+}_{B^{s,m}_{p,q}} \le C\|f\|_{B^{s,m}_{p,q}}$$

follows directly from Theorem 2 (observe that the integral in the basic definition of the Besov space semi-norm may be discretized by using the monotonicity properties of the moduli of smoothness with respect to the sequence $t = t_j \equiv a^{-j}$, $j \ge 0$, $t_{-1} = \infty$ such that

$$\|f\|_{B^{s,m}_{p,q}} \approx \|f\|_p + \|\{a^{js}\omega(a^{-j}, f)_p\}\|_{l_q}$$

becomes obvious).

Thus, it remains to show that

$$\|f\|_{B^{s,m}_{p,q}} \le C\|f\|^{*}_{B^{s,m}_{p,q}}$$

This can be seen directly by applying the idea of the proof of Theorem 5. To this end, let $\|f\|^{*}_{B^{s,m}_{p,q}}$ be finite, and consider an arbitrary admissible decomposition of $f \in L_p(\Omega)$

$$f = \sum_{j=0}^{\infty} g_j \quad , \qquad \|\{a^{sj}\|g_j\|_p\}\|_{l_q} < \infty$$

(the L_p-convergence of this decomposition is obvious by these assumptions). Now we estimate for $1 \le q < \infty$ by standard L_p - l_q inequalities (Minkowski, Hölder, etc.) and by the Bernstein-type inequality of Theorem 4 (the justification of the convergence of the infinite series appearing below as well as the case $q = \infty$ are left as an exercise):

$$|f|^{q}_{B^{s,m}_{p,q}} \le C \left\{ \left(\sum_{j=0}^{\infty} \|g_j\|_p \right)^q + \int_0^1 \left(\sum_{j=0}^{\infty} \omega_m(t, g_j)_p \right)^q t^{-sq-1}\, dt \right\}$$

$$\le C \left\{ \sum_{j=0}^{\infty} a^{\epsilon qj} \|g_j\|_p^q \right.$$

$$\left. + \sum_{l=1}^{\infty} \int_{t_l}^{t_{l-1}} \left(\sum_{j=0}^{\infty} \omega_m(t, g_j)_p \right)^q t^{-sq-1}\, dt \right\}$$

$$\le C \left\{ \sum_{j=0}^{\infty} a^{\epsilon qj} \|g_j\|_p^q \right.$$

$$\left. + \sum_{l=1}^{\infty} a^{sql} \left(\sum_{j=0}^{l-1} a^{\lambda(j-l)} \|g_j\|_p + \sum_{j=l}^{\infty} \|g_j\|_p \right)^q \right\}$$

$$
\leq \; C\left\{\sum_{j=0}^{\infty} a^{\epsilon q j}\|g_j\|_p^q + \sum_{l=1}^{\infty} a^{sql}\cdot\right.
$$

$$
\left.\left(\sum_{j=0}^{l-1} a^{(\lambda-\epsilon)q(j-l)}\|g_j\|_p^q + a^{-\epsilon ql}\sum_{j=l}^{\infty} a^{\epsilon q j}\|g_j\|_p^q\right)\right\}
$$

$$
\leq \; C\|\{a^{sj}\|g_j\|_p\}\|_{l_q}^q \; .
$$

This holds whenever ϵ has been chosen from the non-empty interval $(0, \lambda - s)$ (the last step in the estimation procedure is simply a change of the summation order). Taking the infimum with respect to all such representations we see the required estimate. The proof of Theorem 6 is complete.

Remark. The conditions of this theorem are relatively restrictive and not immediately fulfilled for many f.e. approximation schemes (especially the monotonicity condition is often violated). Nevertheless, they hold for some typical f.e. constructions on dyadically refined triangulations or quadrilateral partitions such as: Lagrange C^0-elements [Os1, Os4] (see also [BY] for a more detailed discussion for linear elements), certain types of composite C^1 elements [Os3, DOS], the Bogner-Fox-Schmit element [Os8] (in all these cases, $a = 2$ is the appropriate number). Moreover, these cases may serve as basis for further modifications applicable to more complicated situations. We come back to this point in section 4. The incorporation of boundary conditions into the approximation schemes which is necessary for dealing with essential boundary conditions for elliptic problems will be discussed in section 3.

For certain classes of splines and shift-invariant approximation schemes, results of this type are given in [DP2, Os2, FJ, FJW, DJP, DK]. The equivalent norm $\|f\|^*_{B^{s,m}_{p,q}}$ will play an important role for analyzing multilevel methods for symmetric variational problems in Sobolev spaces $(p = q = 2)$ which will be pointed out in 4.2.

3 Function spaces

The idea of this section is to give some general information about the theory
of function spaces in the spirit of the "decomposition approach" which goes
back to the famous Littlewood-Paley theorem for trigonometric Fourier series
of L_p-functions on the one hand (for $p = 2$ all this simply reduces to the Bessel
identity), and to the direct and inverse inequalities of approximation theory by
trigonometric polynomials (cf. the previous section for the piecewise polynomial
counterpart) on the other. We follow the philosophy of [Tr2, Tr3, Tr4] : first
spaces on $\mathbf{R}^d$ have to be studied carefully, then one canonically defines spaces
on domains by restriction, i.e. if $X \equiv X(\mathbf{R}^d)$ is some normed space of functions
(distributions) defined on $\mathbf{R}^d$ then

$$X(\Omega) := \{g \mid g = f|_\Omega \text{ for some } f \in X\} , \quad \|g\|_{X(\Omega)} = \inf_{g=f|_\Omega} \|f\|_X ,$$

and checks whether this makes sense. This last step of the procedure strongly
depends on the solution of the extension problem for spaces given by intrinsic
properties (examples are the Sobolev classes for $m = 1, 2, \ldots$ and $1 \le p \le \infty$ and
the Besov spaces of positive smoothness which are directly defined as classes of
functions on Ω without referring to any kind of behavior outside the domain,
see [EE, Gv, Nč]). These things are briefly discussed in subsections 3.1 and
3.2. The trace problem which is important for the correct understanding of
boundary value problems (but also for things like boundary integral equations,
the boundary element and domain decomposition methods) will be touched on
in 3.3.

However, the main subsection is 3.4 where we develop the concept we started
with in 2.6 : For reasonable sequences of finite element subspaces

$$V_0 \subset V_1 \subset \ldots \subset V_j \equiv V(\mathcal{T}_j) \subset \ldots$$

we study the spaces defined by the decomposition norm $\| \cdot \|^*_{B^{s,m}_{p,q}}$ introduced
in Theorem 6 on their own. It turns out that extension properties, traces,
embeddings etc. for these approximation spaces can be understood in a very elementary way. Further applications follow in the subsequent sections. It

should be mentioned that our approach is close to the concept of so-called local approximation spaces resp. of atomic representations a general theory of which has been developed only recently. See [Tr4, FJW] for more information and historical comments.

3.1   Spaces on $\mathbf{R}^d$

3.1.1   Fourier decomposition methods

We will assume that there is some familiarity with the notion of distributions from the class of tempered distributions $\mathcal{S}' \equiv \mathcal{S}'(\mathbf{R}^d)$ and Fourier transform techniques applied to functions or distributions (e.g., from some L_p or the Schwarz class $\mathcal{S}$ or from $\mathcal{S}'$). If one is only interested in understanding what happens in the applications, one may think about arbitrary L_2 functions. For those, the Fourier transform is given by

$$\mathcal{F}f(\xi) = \text{v.p.} \, (2\pi)^{-d/2} \int_{\mathbf{R}^d} f(x)e^{-ix\cdot\xi}\,dx$$

and the inverse Fourier transform by $\mathcal{F}^{-1}f(x) = \mathcal{F}f(-x)$. Plancherel's theorem states that $\mathcal{F}$ is an isometry of $L_2(\mathbf{R}^d)$ onto itself, the bijectivity holds true for $\mathcal{S}$ as well. It can be extended to all of $\mathcal{S}'$ and is a bijection on this class of distributions.

Using this information and the formulae for the Fourier transform of derivatives, the following statement is clear:

$$f \in W_2^m(\mathbf{R}^d) \iff (1+|\xi|^2)^{m/2}|\mathcal{F}f(\xi)| \in L_2(\mathbf{R}^d)\,.$$

Such expressions also appear if one studies the mapping properties of the Laplacian $-\Delta$ and its (fractional) powers. This led to the definition of the classes

$$H_p^s(\mathbf{R}^d) = \{f \in \mathcal{S}' \,|\, \|f\|_{H_p^s} = \|\mathcal{F}^{-1}((1+|\xi|^2)^{s/2}\mathcal{F}f(\xi))\|_p < \infty\}\,,$$

$1 < p < \infty$, $s \in \mathbf{R}$, which go back to Hörmander ($p = 2$) and are usually called Bessel potential spaces. As stated above, we have $H_2^m = W_2^m$ for integer $s = m$ (cf. Theorem 7 below for other p), i.e. the spaces H_p^s may be viewed as a continuation of the Sobolev spaces to non-integer smoothness parameters [LM, Gv, EE]. Another extension was created during the fifties in connection with the trace problem: For noninteger $s > 0$, and $1 \le p < \infty$ Slobodetzkij defined W_p^s to be the class $B_{p,p}^{s,m}$ ($m > s$, the actual definition was given in terms of first and second differences of the $[s]$-th derivatives of functions). Thus, the Slobodetzkij classes are a forerunner of the concept of Besov spaces.

The attempt to properly generalize this definition to all values $p \neq 2$ and to cover both the definition of Sobolev spaces and of Besov spaces led to the following concept which is mainly due to Peetre, Lizorkin, and Triebel (see the historical comments in [Tr4]): Let $\{\phi_j(\xi)\}$ form a partition of unity on $\mathbf{R}^d$ generated by a single C^∞ function $\phi_0(\xi)$ with value 1 for $|\xi| < 1$ and vanishing for $|\xi| > a$ by the formulae $\phi_j(\xi) = \phi_0(a^{-j}\xi) - \phi_0(a^{-j+1}\xi)$ for all $j > 0$. The number $a > 1$ will be fixed (it does not matter for the function classes to be defined, different $a > 1$ lead to equivalent norms). It is clear that the support of ϕ_j is contained in the set $\{a^{j-1} \leq |\xi| \leq a^{j+1}\}$ if $j > 0$. In [Tr2, Tr3, Tr4] more general choices for this partition of unity are discussed.

Definition. A distribution $f \in \mathcal{S}'$ belongs to $F^s_{p,q} \equiv F^s_{p,q}(\mathbf{R}^d)$ resp. to $B^s_{p,q} \equiv B^s_{p,q}(\mathbf{R}^d)$ iff

$$\|f\|_{F^s_{p,q}} = \| \{a^j \mathcal{F}^{-1}(\phi_j \mathcal{F} f)\}_{j\geq 0} \|_{L_p(l_q)} < \infty$$

resp. iff

$$\|f\|_{B^s_{p,q}} = \| \{a^j \mathcal{F}^{-1}(\phi_j \mathcal{F} f)\}_{j\geq 0} \|_{l_q(L_p)} < \infty \quad .$$

It is easy to check that the F-spaces and the B-spaces coincide for $p = q$, moreover, if $p = q = 2$ then $H^s \equiv H^s_2$ comes out (with equivalent norms). The following theorem holds true.

Theorem 7. The scales F and B consist of Banach spaces for all parameters $1 \leq p, q \leq \infty$ and $s \in (-\infty, \infty)$. Different partitions of unity (defined by ϕ_0 and $a > 1$) lead to the same spaces, with equivalent norms. The following concrete spaces are particular cases of these definitions ($\Omega = \mathbf{R}^d$):

- Sobolev spaces : $W^m_p = F^m_{p,2}$, $1 < p < \infty$, $m = 0, 1, \ldots$,

- Bessel-potential spaces : $H^s_p = F^s_{p,2}$, $1 < p < \infty$, $-\infty < s < \infty$,

- Besov-Lipschitz spaces : $B^{s,m}_{p,q} = B^s_{p,q}$, $0 < s < m$.

We sketch the elementary proof of the norm equivalence for the Besov spaces with $p = q = 2$ which will be needed in the applications of section 4. Another approach serving all mentioned parameters is the approximation method: one can prove direct and inverse theorems, this time for the best approximations with respect to entire analytic functions of exponential type (with Fourier transforms supported in a ball), and obtain characterizations of the Besov spaces in the same manner as described in 2.6 for the finite element case, see [Ni].

Since for $p = q = 2$ the new B- and F-norms coincide and can be expressed in terms of the Fourier transform, we want to look for a Fourier transform argument. Obviously,

$$\mathcal{F}(\Delta_h^m f)(\xi) = \sum_{l=0}^{m}(-1)^{m-l}\binom{m}{l} e^{ilh\cdot\xi}\mathcal{F}f(\xi) = (1 - e^{ih\cdot\xi})^m\mathcal{F}f(\xi)$$

which immediately gives $(|1 - e^{it}| = 2|\sin(t/2)|$!)

$$I_m(h, f)_2^2 = \int_{\mathbf{R}^d}(2\sin(h\cdot\xi/2))^{2m}|\mathcal{F}f(\xi)|^2\,d\xi\;.$$

Integrating with respect to h over the ball $B(0,t)$, and stressing some knowledge in calculus we obtain the two-sided inequalities

$$\omega_m(t, f)_2^2 \approx t^{-d}\int_{B(0,t)}I_m(h, f)_2^2\,dh \approx \int_{\mathbf{R}^d}\min\left((|\xi|t)^{2m}, 1\right)|\mathcal{F}f(\xi)|^2\,d\xi$$

Now it remains to multiply by t^{-2s-1} and to integrate with resect to $t \in (0,\infty)$ (have in mind that $0 < s < m$). After adding the term $\|f\|_2^2 = \|\mathcal{F}f\|_2^2$ one arrives at

$$\|f\|_{B_{2,2}^{s,m}}^2 \approx \int_{\mathbf{R}^d}(1 + |\xi|^{2s})|\mathcal{F}f(\xi)|^2\,d\xi\;\;.$$

The latter expression is obviously equivalent to that given in the Besov norm definition (use that $\{\phi_j\}$ forms a partition of unity with the support properties mentioned above).

3.1.2   Other techniques and spaces

We do not want to give too many details. We only give a list of possibilities how to rewrite the above definitions in other terms (a nice and comprehensive overview is given in the introductory part of the recent book [Tr4], cf. also [BKL, Tr5]):

- Definitions by *derivatives* and *differences* : Instead of the moduli of smoothness of f one may use directly $I_m(h, f)_p$ (or combinations of $I_k(h, \frac{\partial^\alpha f}{\partial x^\alpha})_p$) to define Besov classes, certain characterizations in terms of local properties of differences can be given also for the F spaces, see [Tr4].

- *Representations by entire analytic functions of exponential type* : Besides characterizations by best approximations, the following representation theorem holds true [Ni] : For $s > 0$, $1 \leq p, q \leq \infty$, f belongs to $B_{p,q}^s$ iff there exists a

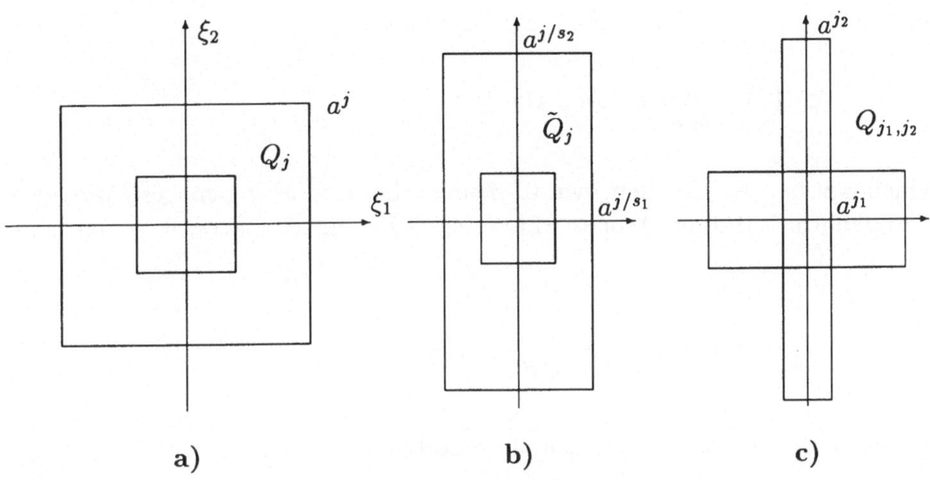

Figure 7.

representation

$$f = \sum_{j=0}^{\infty} g_j \qquad \text{(convergence in } \mathcal{S}'\text{)} \qquad \text{supp } \mathcal{F}g_j \subset Q_j = B(0, a^j)$$

into entire analytic functions of exponential type $\leq a^j$ with the following property controlling the growth of the terms in the representation:

$$\|\{a^{js}g_j(\cdot)\}\|_{l_q(L_p)} < \infty$$

The infimum over these expressions with respect to all representations yields an equivalent norm in $B_{p,q}^s$. Instead of balls, one might use cubes Q_j of side-length a^j centered at the origin in the above definition, see Figure 7 a). Analogous statements hold for F-spaces, with $l_q(L_p)$ replaced by $L_p(l_q)$, see [Ni, Tr2, Tr4].

- *Spectral representations* (like those arising in connection with $-\Delta$, see [LM, Tr2]).

- *Local polynomial approximation* and *oscillations*: The observation is that averages of differences (such as moduli of smoothness) can be substituted locally by means of best polynomial approximation over small balls, compare the Whitney-type inequality stated in 2.3 (Theorem 3). The important quantities are local oscillations

$$\text{osc}_p^m f(x,t) = \inf_{p_m \in \mathbf{P}_m} |B(x,t)|^{-1/p}\|f - p_m\|_{L_p(B(x,t))}$$

and corresponding (sharp) maximal functions

$$f_p^{m,s}(x) = \sup_{0<t<1} t^{-s}\mathrm{osc}_p^m f(x,t)$$

which can be used on their own to define other interesting spaces (Morrey - Campanato, BMO, etc.) or to characterize F- and B-spaces. E.g., one has $f \in B_{p,q}^s$ iff

$$\|f\|_p + \left(\int_0^1 \|\mathrm{osc}_p^m f(\cdot,t)\|_p^q t^{-sq-1}\, dt\right)^{1/q} < \infty$$

for $0 < s < m$ (for $q = \infty$ one should modify correspondingly).

- *Atomic, spline, and wavelet decompositions.* The historical roots are the Haar system (1910), the investigations on spline systems in function spaces on the one side, and the theory of real Hardy spaces developed in the seventies, on the other. Over the last few years, we have seen a considerable activity in using atoms (simple, localized building blocks) for representation and approximation purposes. Signal analysis is one applied field that stimulates these activities. We refer to [Tr4], section 1.9, for definitions of atoms and atomic characterizations of function spaces, and to the wavelet literature in general for the more applied aspect. One should bear in mind that finite element schemes are conceptionally atomic: The nodal basis functions are, properly normalized, atoms with smoothness of low order.

Another important tool to work in a scale of function spaces which we will just mention is the *interpolation theory*. The classical background is the following observation: if some result, say, the boundedness of a linear operator, holds for L_{p_0} and for a second L_{p_1} ($1 \leq p_0 < p_1 \leq \infty$) then it remains true for all L_p spaces with $p_0 < p < p_1$, see [BL, Tr2] for extensive information.

The above spaces are *isotropic* and *non-weighted* by definition, and, therefore, are designed for dealing with nondegenerating elliptic boundary value problems with smooth coefficients on smooth domains (here even on $\mathbf{R}^d$). For problems that involve, e.g., the differential operator $-\nabla(a\nabla u)$ with strongly varying or degenerating coefficient function $a(\cdot) \geq 0$ one might be interested in *weighted spaces* over domains, these are useful also in connection with the singularity problem for elliptic equations in non-smooth domains ([Gv, Dg, KS]). The Fourier analytic approach to these spaces is discussed in [Tr2]. *Anisotropic spaces* allow for direction-depending smoothness control. Anisotropic Besov-Sobolev spaces are dealt with in [Ni, BIN, BKL], their definition can be obtained formally from the above ones by introducing Sobolev semi-norms like

$$|f|_{p,\mathbf{m}} = \sum_{i=1}^{d} \|\frac{\partial^{m_i} f}{\partial x_i^{m_i}}\|_p \qquad (\mathbf{m} = (m_1, \dots, m_d))$$

(anisotropic Sobolev spaces), using directional moduli of smoothness as defined in 2.2 (anisotropic Besov spaces) or allowing more general partitions of unity, with supporting sets others than $\{x \,|\, a^{j-1} \leq |x| \leq a^{j+1}\}$, and more general smoothness control given by sequences others than $\{a^j\}$ (Fourier transform definition). If one looks again for charcterizations via representations by entire analytic functions of exponential type then the right condition is (we take once again the Besov scale as example)

$$\|f\|_{B_{p,q}^s} = \inf_{f=\sum g_j} \|\{a^j g_j\}\|_{l_q(L_p)} < \infty \quad ,$$

where

$$\operatorname{supp} \mathcal{F} g_j \subset \tilde{Q}_j(\mathbf{s}) = \{x \in \mathbf{R}^d \,|\, |x_i| < a^{j/s_i}, \, i = 1, \dots, d\} \quad , \quad j \geq 0 \,.$$

The smoothness vector $\mathbf{s} = (s_1, \dots, s_d)$ enters the support control of the g_j, cf. Figure 7 b). For $s_i = s = \text{const.} > 0$, the spaces coincide with their isotropic counterparts defined above.

Finally, in some problems, spaces of functions with *dominating mixed derivative* are of interest, see [BKL], section 3.3, or the recent book [Tl] for classical definitions, and [ST] for a theory in the spirit of the Fourier decomposition approach. The right version of representations by entire analytic functions is now to work with multiple series $f = \sum_{\mathbf{j}} g_{\mathbf{j}}$, $\mathbf{j} \in \mathbf{Z}^d$, where the supports of the $\mathcal{F} g_{\mathbf{j}}$ are in the sets $Q_{\mathbf{j}}$ indicated in Figure 7 c), if one takes the factors $a^{\mathbf{j} \cdot \mathbf{s}}$ in the norm definition then spaces with dominating mixed s-th derivative come out. For instance, the corresponding F-space definition for the case $\mathbf{s} = (2, 2)$ would lead to the mixed Sobolev space $(1 < p < \infty)$

$$S_p^{(2,2)} W = \{f \in L_p \,|\, \|f\|_p + \|\frac{\partial^2 f}{\partial x_1^2}\|_p + \|\frac{\partial^2 f}{\partial x_2^2}\|_p + \|\frac{\partial^4 f}{\partial x_1^2 \partial x_2^2}\|_p < \infty\} \,.$$

We come back to such classes in 4.5.

Note that besides the classical Sobolev spaces and the less used but still standard Besov spaces [BA, Va], some of the above mentioned function classes and methods have already found admiration in numerical analysis: BMO (the

John-Nirenberg class of functions of bounded mean oscillation) and the Morrey-Campanato spaces were used to understand better the problems around L_∞ estimates for finite element projections, weighted spaces and norms to deal with the pollution effect for corner points, various function classes of mixed type are used for error estimates for blending approximation schemes, in the theory of optimal cubature and recovery.

3.2   Spaces on domains and extension

We recall that for domains we have now at least two ways to create definitions for various function spaces: in all cases one could use the anonymous restriction trick mentioned at the very beginning of section 3. Some of the above, on $\mathbf{R}^d$ equivalent definitions can be localized to domains directly. This is the case with the original definition of Sobolev spaces W_p^s and Besov spaces $B_{p,q}^{s,m}$ but local polynomial approximation, atomic representations, even spectral methods will do as well. Such definitions are called intrinsic.

There is a certain hope that both possibilities lead to the same result. This can be proven by means of the so-called restriction and extension procedure. As long as our spaces are spaces of functions resp. distributions, and our domains Ω open subsets of $\mathbf{R}^d$, the restriction operator is simply

$$\mathcal{R} : f \to f|_\Omega \quad ,$$

and Ω has the *restriction property* with respect to a certain space $X(\cdot)$ (intrinsic definition) if

$$\|f|_\Omega\|_{X(\Omega)} \le C\|f\|_{X(\mathbf{R}^d)} \qquad \forall\, f \in X(\mathbf{R}^d) \,.$$

Usually, this is trivially satisfied for all domains.

The opposite direction is the *extension property*. A domain Ω has the extension property for a given space $X(\cdot)$ if there exists a mapping $\mathcal{E} : X(\Omega) \to X(\mathbf{R}^d)$ such that $\mathcal{R}\mathcal{E} = \mathrm{Id}$ and

$$\|f\|_{X(\mathbf{R}^d)} \le C\|f|_\Omega\|_{X(\Omega)} \qquad \forall\, f \in X(\Omega) \,.$$

If this extension mapping is linear, things are much easier but even nonlinear extension operators are allowed.

It turns out that extension is sometimes impossible. E.g., take a slit domain and the space of continuous functions $C(\bar{\Omega})$. The class of domains where extension is possible (for almost all types of function spaces used in life) are Lipschitz graph domains or domains satisfying some uniform cone condition or minimally smooth domains (in the sense of Stein [St]). There is some non-uniformity in the literature concerning these almost equivalent properties. Since we are restricted to bounded polyhedral domains, the uniform cone condition given after Theorem 1 in 2.2 will do the job. For some comparison of different properties of domains and their boundaries, see also [Ad, Sp, Gv]. To give a feeling, and in order to derive at least one result we need in the sequel, we sketch the two-three main known extension procedures for classes of smooth functions (cf. [St, Bu, BG]).

(i) *Hestenes extension.* This is elementary and one-dimensional at the first glance : Take some real numbers $0 < \lambda_1 < \ldots < \lambda_{N+1}$, and define the a_k's from the conditions

$$\sum_{k=1}^{N+1} a_k(-\lambda_k)^l = 1 \ , \qquad l = 0, \ldots, N \ .$$

Such numbers exist uniquely. Then for a univariate function defined on the half-line $x \geq 0$, we set

$$\mathcal{E}_{Hest,N} f(x) = \begin{cases} f(x) , & x \geq 0 \\ \\ \sum_{k=1}^{N+1} a_k f(-\lambda_k x) , & x < 0 \end{cases}$$

By construction it is clear that this linear extension operator preserves univariate polynomials given for $x \geq 0$ up to degree N (for $N = 0$ we simply get even extension). This in turn ensures that $\mathcal{E}_{Hest,N}$ can be used for extension of various function classes with smoothness order $\leq N$ or even $\leq N+1$ from $[0, \infty)$ to the whole line. In the multivariate case, this is the step from the half-space to the whole $\mathbf{R}^d$, cf. [Tr4] where some generalization of the Hestenes procedure is introduced. In principle, Hestenes extension is well-suited also for polyhedral domains. However, proofs get complicated even with this simple procedure near corner points and edges.

(ii) *Whitney extension.* This extension procedure was designed (and used) for extending functions satisfying Lipschitz conditions from an arbitrary closed set F to $\mathbf{R}^d$. It relies on the Whitney decomposition of the open complement $D = \mathbf{R}^d \backslash F$ into dyadic cubes $\{Q_k\}$. Look at Figure 8. Roughly speaking, the cubes should have pairwise disjoint interiors and satisfy the condition $\mathrm{dist}(Q_k, F) \approx l(Q_k)$ with proper constants for all k, with $l(Q_k)$ the side-length of the cube which equals some integer power of 2. Then one can construct a C^∞ partition of unity $\{\psi_k\}$ on D such that

- supp ψ_k is located in a cube $\tilde{Q}_k \subset D$ which has the same center as Q_k but is expanded by some constant factor.

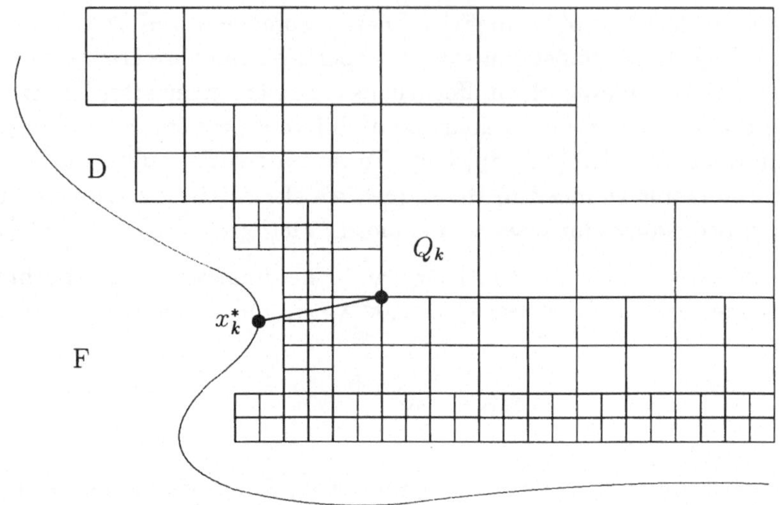

Figure 8. Whitney decomposition

- $|D^m \psi_k| \leq C l(Q_k)^{-m}$, $m \geq 0$.

- $\sum_k \psi_k(x) \equiv 1$ on D .

- At each point $x \in D$ only a finite number of ψ_k does not vanish.

This construction which is described in detail in [St] also yields the existence of a C^∞ function $\delta(x)$ defined on D such that $\delta(x) \approx \mathrm{dist}(x, F)$ for $x \in D$ (the so-called regularized distance to F).

The hierarchy of Whitney extension operators is defined by

$$\mathcal{E}_{Whit,N} f(x) = \sum_k p_{N+1}(x; x_k^*, f)\psi_k(x) , \quad x \in D$$

where $x_k^* \in F$ is any fixed point where the minimal distance between Q_k and F is attained. The summation is extended over all cubes of sidelength ≤ 1, and p_{N+1} denotes the N-th degree Taylor polynom of f expanded at the point x_k^* (this makes sense if f is N-times differentiable on F, a description what this means for functions on an arbitrary closed set F as well as the definition of the Lipschitz class $C^{N,\alpha}(F)$ are given in [St]).

(iii) *Stein extension.* This is in some sense a refinement of both (i) and (ii). Consider $d = 2$. For a single Lipschitz graph domain

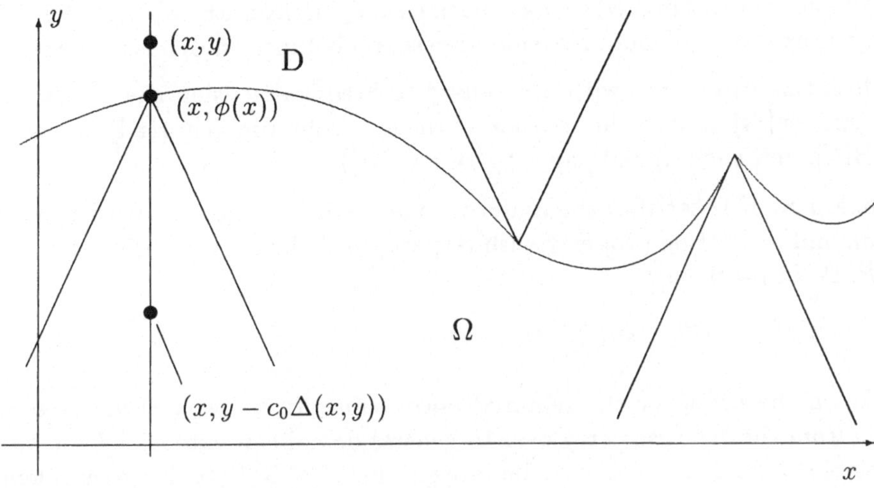

Figure 9. Stein extension for Lipschitz graph domains

$$\Omega = \{(x,y) \in \mathbf{R}^2 \mid y < \phi(x)\,,\ -\infty < x < \infty\}$$

given by a globally Lipschitz continuous function ϕ we set $F = \bar{\Omega}$ (cf. Figure 9, note that such a domain satisfies the uniform cone condition with one set $U = V = \mathbf{R}^d$ and one (infinite) cone L with axis parallel to the y-axis and an angle depending on the global Lipschitz constant of ϕ), and

$$\mathcal{E}_{Stein} f(x,y) = \int_1^\infty f(x,y - c_0\lambda\Delta(x,y))a(\lambda)\,d\lambda\,,\ y > \phi(x)\,,\ x \in \mathbf{R}\,.$$

Here, $c_0 > 0$ will (and can) be chosen such that the integration point is properly inside Ω, and $a(\cdot)$ is rapidly decreasing and continuous on $[1,\infty)$ with

$$\int_1^\infty a(\lambda)\lambda^k\,d\lambda = \begin{cases} 1\,, & k = 0 \\ 0\,, & k > 0 \end{cases} \quad \forall\, k\,.$$

The existence of such a function which substitutes the finite sequence $\{a_k\}$ from (i) is shown in [St]. The point is that $\mathcal{E}_{Stein}$ can be used to define an extension operator for domains satisfying the uniform cone condition (construct first a C^∞ partition of unity on the basis of V_i, U_i, cut an arbitrary function into a finite number of pieces concentrated on $U_i \cup \Omega$, and apply to each piece the (rotated) $\mathcal{E}_{Stein}$!). This solves the extension problem for Sobolev spaces W_p^m with $1 \leq p \leq \infty$ and $m = 1, 2, \ldots$ once at all (there is one more method due to A.P.Calderon which did this before for $1 < p < \infty$, we refer to the references).

Modifications of this procedure have been used by many authors, e.g. [JS2, BIN] where the extension of moduli of smoothness and K-functionals has been treated.

We close this subsection with a corollary to Stein's extension theorem for Sobolev spaces [St], and to the solution of the extension problem for Besov spaces (cf. [BIN], and the original paper by Besov [Be]).

Theorem 8. If Ω satisfies the uniform cone condition, then it satisfies the extension and restriction property with respect to all classes W_p^s, H_p^s $(1 < p < \infty)$ and $B_{p,q}^{s,m}$. In particular,

$$H^s(\Omega) = B_{2,2}^{s,m}(\Omega) , \qquad 0 < s < m .$$

We sketch the proof for the required norm equivalence (we will also see how results from the $\mathbf{R}^d$ case carry over to general domains if extension/restriction is possible). Let $0 < s = k < m$ be integer. Let $f \in B_{2,2}^{k,m}(\Omega)$ be given, denote by $\tilde{f}$ its extension suitable for Besov spaces (use for instance the construction given in [JS2]). Thus, by Theorem 7, and the explicit definition of H^k

$$\|f\|_{H^k(\Omega)} \leq \|\tilde{f}\|_{H^k(\mathbf{R}^d)} \leq C\|\tilde{f}\|_{B_{2,2}^{k,m}(\mathbf{R}^d)} \leq C\|f\|_{B_{2,2}^{k,m}(\Omega)}$$

The opposite direction is analogous, this time we use the Stein extension operator under the assumption $f \in H^k(\Omega) = W_2^k(\Omega)$ in order to switch to $\mathbf{R}^d$, and so on. The non-integer case can be dealt with by interpolation.

3.3   Spaces on manifolds and traces

We will be very brief. The problems of the existence of traces onto manifolds of functions from different classes, and the construction of spaces on them have the same origin in the solution theory of boundary value problems for partial differential equations. We are interested in functions from classes used in the variational theory - the Sobolev classes and related function spaces. These are subspaces of L_p, i.e. functions in these spaces are a priori defined only up to a set of zero measure in $\mathbf{R}^d$. Thus, if we look at a manifold Γ of dimension $< d$ in Ω then we see the problem. On the other hand, for very smooth functions restrictions are well-defined on any subset of Ω. Since smooth functions form, as a rule, dense subsets in most of our function classes we see the solution: If $X(\Omega)$ is a function space, and Γ a submanifold of Ω then we suppose that there exists a mapping tr_Γ defined on a dense subset of $X(\Omega)$ with $tr_\Gamma(f)$ being some function on Γ. Then

$$X(\Gamma; tr_\Gamma) = \overline{\{h = tr_\Gamma(f) \mid f \in \mathrm{dom}(tr_\Gamma) \cap X(\Omega)\}}$$

where the closure is taken with respect to the factor norm

$$\|h\|_{X(\Gamma, tr_\Gamma)} = \inf_{f \in X(\Omega) : h = tr_\Gamma(f)} \|f\|_{X(\Omega)} .$$

If the trace operator represents the usual restriction of functions defined on Ω onto Γ we also use the the familiar notation $|_\Gamma$ instead of tr_Γ and $X(\Omega)|_\Gamma$ for the spaces. One can clearly consider traces like restrictions of (normal) derivatives etc..

After having a general definition one may ask for concrete descriptions. This characterization problem for trace spaces (and the correct understanding of the mapping theory for differential operators on domains including the differential operators prescribing the set of boundary conditions) is one of the hardest problems in the field. Besides some historical problems (traces of Sobolev spaces with $p \neq 2$ are not Sobolev spaces with respect to the manifold but Besov spaces) which are now overcome, the difficulties arise since the geometry of the manifold comes in.

Basic facts on trace spaces can be found in [LM, Tr2], the case of a polygonal and other domains with piecewise smooth boundary is considered in [Gv]. We will come back to traces onto piecewise polyhedral manifolds in the next section (3.4.4) and consider the construction of the trace classes for approximation spaces with respect to multilevel schemes of finite element spaces. The immediate applications (domain decomposition methods) are touched on in 4.4.

3.4   Approximation spaces on polyhedral domains

This is the main subsection of this chapter. We come back to sequences of f.e. subspaces $\{V_j\}$ over a polyhedral domain $\Omega \in \mathbf{R}^d$ as studied in section 2, and investigate a scale of approximation spaces $A^s_{p,q}(\{V_j\})$ which implicitly appeared in Theorem 6 and was introduced in [Os1]. It might be viewed as a (much richer) particular case of the general theory of approximation spaces developed in the late seventies [PS, BK, BS, Pi] as well as a variant of spline characterization theorems for Besov-Lipschitz spaces obtained by several authors, and already mentioned before. Nevertheless, we can still add some epsilons (trace and extension theorems), and will be able to address some new applications to the analysis of multilevel f.e. methods in the next sections.

Throughout this section, we will assume that the monotonicity condition

(M) $\qquad V_0 \subset V_1 \subset \ldots \subset V_j = V(\mathcal{T}_j) \subset \ldots$

is fulfilled, that the partitions $\{\mathcal{T}_j\}$ are regular and quasi-uniform, with $h(\mathcal{T}_j) \approx$

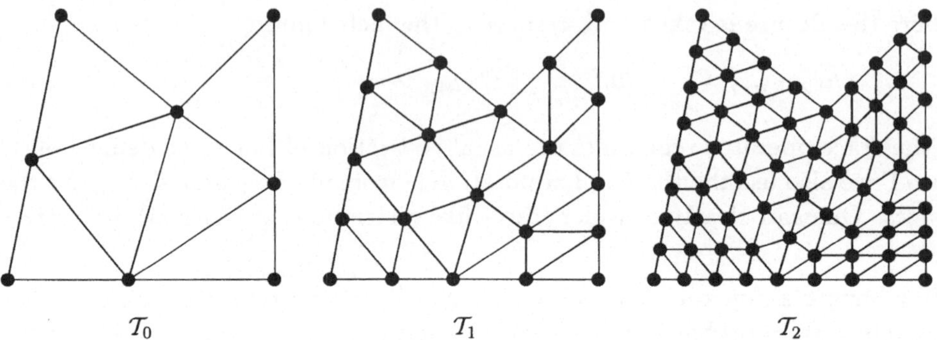

Figure 10. Regular dyadic refinement

$h_{\min}(\mathcal{T}_j) \approx a^{-j}$ for some fixed $a > 1$, and that **(A1)** holds (due to the assumed quasi-uniformity, the latter simplifies to

$$\textbf{(A1)'} \qquad \|g_j\|_p \approx a^{-jd/p}\|\{c_{j,i}\}\|_{l_p} \quad \forall g_j = \sum_i c_{j,i} N_{j,i} \in V_j \, , \, j \geq 0 \, , \, 1 \leq p \leq \infty$$

which also explains the notations for the nodal basis functions and coefficients of f.e. functions from different V_j used below). The standard and most simple example one can use to better understand the exposition is the following: Start with a regular triangulation $\mathcal{T}_0$ of a polyhedral domain in $\mathbf{R}^2$ or $\mathbf{R}^3$, produce $\{\mathcal{T}_j\}$ successively by regular dyadic refinement (subdivision of each triangle resp. tetrahedron into 4 new triangles resp. 8 tetrahedra of exactly (or approximately) the same size and shape) such that the triangulations remain regular and quasi-uniform with some constants γ_0 and γ_1, and take the subspaces of linear f.e. functions as $V(\mathcal{T}_j)$, cf. Figure 10. In this case all assumptions are satisfied with $a = 2$. Note that any other f.e. scheme of the (small) list mentioned at the end of 2.6 can be used as well.

3.4.1 Definition and general properties

Let $1 \leq p \leq \infty$, $s > 0$, $1 \leq q \leq \infty$ or $s = 0$, $q = 1$. We say that $f \in L_p(\Omega)$ belongs to $A_{p,q}^s(V_j)$ if there exists an *admissible representation*

$$\textbf{(R)} \qquad f = \sum_{j=0}^{\infty} g_j \, , \quad g_j \in V_j \quad (j = 0, 1, \dots) \quad \text{with} \quad \|\{a^{sj}g_j(\cdot)\}\|_{l_q(L_p)} < \infty \, .$$

Furthermore, set

$$\|f\|_{A_{p,q}^s} \doteq \inf_{\textbf{(R)}} \|\{a^{sj}g_j(\cdot)\}\|_{l_q(L_p)} \quad .$$

A few comments are necessary. For all parameters, the above admissible representations converge, at least in the sense of L_p. Thus, we get non-empty linear subspaces of L_p:

$$\cup_{j=0}^{\infty} V_j \subset A_{p,q}^s(\{V_j\}) \subset \hat{L}_p = \overline{\cup_{j=0}^{\infty} V_j}\Big|_{L_p} \subset L_p$$

As a rule we have $\hat{L}_p = L_p$ for $1 \le p < \infty$ and $\hat{L}_\infty = C$ (this holds for the above-mentioned f.e. types). Moreover,

$$A_{p,1}^0(\{V_j\}) = \hat{L}_p \quad , \qquad \|f\|_{A_{p,1}^0} = \|f\|_p \quad .$$

This is a simple consequence of the Minkowski inequality and the definition of $\hat{L}_p$.

For $s > 0$, we can compare with Theorem 6 of 2.6: under the additional assumptions made there we have equivalences to Besov spaces (and to Sobolev spaces H^s if $p = q = 2$, cf. Theorem 8 in 3.2). Thus, our $A_{p,q}^s$ scale seems to be interesting enough to be investigated. Spaces with other parameter combinations (say, with $s < 0$) do not make sense, at least in the above function setting.

Theorem 9. (Equivalent norms) The above defined spaces are Banach spaces. For $s > 0$ the following expressions form equivalent norms:

$$\|f\|_{A_{p,q}^s}^+ = \|f\|_p + \|\{a^{sj} E_{V_j}(f)_p\}\|_{l_q}$$

$$\|f\|_{A_{p,q}^s}^{**} = \inf_{(\mathbf{R})} \|\{a^{(s-d/p)j}\|\{c_{j,i}\}\|_{l_p}\}\|_{l_q}$$

$$\|f\|_{A_{p,q}^s}^{***} = \|\{a^{sj}(R_j f - R_{j-1} f)\}\|_{l_q(L_p)}$$

for any sequence of projections $R_j : L_p \to V_j$ with uniformly bounded norms $\|R_j\|_{L_p \to L_p}$ ($R_{-1} f \equiv 0$).

Proof. Abstract versions of this theorem, together with statements on interpolation of these spaces can be found in [Pi] and the references cited therein.

That all expressions are norms, follows from the corresponding properties of L_p and l_q norms. The proof of the norm equivalences is the same as that of Theorem 6, with small alterations which are left to the reader. The only thing which remains is the completeness. It suffices to show it with respect to $\|\cdot\|_{A_{p,q}^s}^+$. Let $\{f^{(n)}\}$ be a Cauchy sequence in $A_{p,q}^s(\{V_j\})$. By definition of the $A_{p,q}^s$ spaces it is also a Cauchy sequence in L_p, we therefore may assume $f^{(n)} \to f \in L_p$ in the sense of L_p (if necessary, select a subsequence and change the notation).

Using some obvious properties of the best approximations like

$$E_M(f)_p \leq \|f\|_p \,, \qquad E_M(f+g)_p \leq E_M(f)_p + E_M(f)_p \,,$$

valid for any M, we immediately get

$$E_{V_j}(f^{(m)} - f^{(n)})_p \to E_{V_j}(f - f^{(n)})_p \quad , \qquad m \to \infty \,, \quad n \geq 0 \,.$$

This implies

$$\|f - f^{(n)}\|_p + \|\{a^{sj} E_{V_j}(f - f^{(n)})_p\}_{j \leq k}\|_{l_q}$$

$$= \lim_{m \to \infty} (\|f^{(m)} - f^{(n)}\|_p + \|\{a^{sj} E_{V_j}(f^{(m)} - f^{(n)})_p\}_{j \leq k}\|_{l_q})$$

$$\leq \limsup_{m \to \infty} \|f^{(m)} - f^{(n)}\|_{A_{p,q}^s}$$

for any $k \geq 0$ and $n \geq 0$. Now, let $k \to \infty$ and then $n \to \infty$, by definition of the $A_{p,q}^s$ norm and the Cauchy sequence property it follows that $f - f^{(n)}$ tends to zero with respect to this norm as well. An analogous argument shows $f \in A_{p,q}^s(\{V_j\})$. The proof is completed.

We mention an immediate consequence of **(A1)'** : an imbedding theorem of Sobolev type holds and can be proved within a few lines.

Theorem 10. (Imbedding theorem) For $1 \leq p < p' \leq \infty$ we have the imbedding

$$A_{p,q}^s(\{V_j\}) \subset A_{p',q}^{s-d(1/p-1/p')}(\{V_j\}) \quad , \qquad s > d(1/p - 1/p') \,,$$

with bounded natural injection. Moreover,

$$A_{p,1}^{d(1/p-1/p')}(\{V_j\}) \subset \hat{L}_{p'}(\Omega) \,.$$

To prove this it remains to substitute the following Nikolskij-type inequality into the definition of the $A_{p,q}^s$ norm :

$$\|g_j\|_{p'} \approx a^{-jd/p'} \|\{c_{j,i}\}\|_{l_p'} \leq a^{-jd/p'} \|\{c_{j,i}\}\|_{l_p} \approx a^{jd(1/p-1/p')} \|g_j\|_p$$

holds for all $g_j \in V_j$, $j \geq 0$.

Note that the last embedding may be improved if $p' < \infty$: then the second parameter $q = 1$ can be enlarged to $q = p'$ which is best possible, see [Os6] for the corresponding one-dimensional argument. Other more trivial embeddings, e.g. the fact that

$$A_{p,q}^s(\{V_j\}) \subset A_{p',q'}^{s'}(\{V_j\}) \qquad \forall \, p' \leq p \,, \, q \geq q' \,, \, s \geq s' \,,$$

directly follow from the definition and properties of L_p - l_q norms, and may be combined with Theorem 10 to get all possible imbeddings. These results, in turn, carry over to usual Besov-Sobolev spaces if in addition to the above assumption all requirements of Theorem 2 resp. 4 are satisfied. E.g., if Theorem 2 is applicable then

$$W_p^m(\Omega) \subset B_{p,\infty}^{m,m}(\Omega) \subset A_{p,\infty}^m(\{V_j\}) \quad \text{resp.} \quad B_{p,q}^{s,m}(\Omega) \subset A_{p,q}^s(\{V_j\}) ,$$

$0 < s < m$. If Theorem 4 holds true then we have inverse imbeddings

$$A_{p,q}^s(\{V_j\}) \subset B_{p,q}^{s,m}(\Omega) , \quad 0 < s < \lambda = \min{(r+1+1/p, m)}$$

where m, r are the integers characterizing approximation order and smoothness class of the f.e. subspaces, and $1 \leq q \leq \infty$ is arbitrary. All imbeddings are continuous, the corresponding norm inequalities hold true. These facts will be used whenever we want to rewrite results on $A_{p,q}^s$ in the more familiar language of Besov-Sobolev spaces (direct and inverse theorems for the comparison of $A_{p,q}^s$ and Sobolev spaces will be obtained from those for Besov spaces).

3.4.2 Approximation theory in the $A_{p,q}^s$ scale

Here we quote the following

Theorem 11. (Jackson-Bernstein-type inequalities) Let $0 \leq s < s'$ be given, the other parameters $1 \leq p, q, q' \leq \infty$ be arbitrary (if $s = 0$ then $q = 1$). If we introduce the best approximations

$$E_{V_j}(f)_{A_{p,q}^s} \equiv \inf_{g_j \in V_j} \|f - g_j\|_{A_{p,q}^s}$$

with respect to the A-norms then we have

$$E_{V_j}(f)_{A_{p,q}^s} \leq Ca^{-(s'-s)j} E_{V_j}(f)_{A_{p,q'}^{s'}} \leq Ca^{-(s'-s)j}\|f\|_{A_{p,q'}^{s'}}$$

for all $f \in A_{p,q'}^{s'}(\{V_j\})$, and

$$\|g_j\|_{A_{p,q'}^{s'}} \leq Ca^{(s'-s)j}\|g_j\|_{A_{p,q}^s}$$

for arbitrary $g_j \in V_j$ and $j \geq 0$.

The elementary proof runs as in [Os1], section 3, where a slightly more general version (combining our Theorem 11 with the imbedding theorems) is established.

Interesting to mention is the role of the two different parameters q, q' for establishing results for Sobolev classes: If we look, for instance, for H_p^s, $1 < p < \infty$, estimates for the example of linear f.e. functions over dyadically refined triangulations (with $a = 2$, $r = 0$ and $m = 2$) then using the imbeddings

$$H_p^s(\Omega) \subset B_{p,\infty}^{s,2}(\Omega) \subset A_{p,\infty}^s(\{V_j\}) \quad , \qquad 0 < s \leq 2 \, ,$$

on the one hand, and

$$A_{p,1}^s(\{V_j\}) = B_{p,1}^{s,2}(\Omega) \subset H_p^s(\Omega) \quad , \qquad 0 < s < 1 + 1/p \, ,$$

$$A_{p,1}^0(\{V_j\}) = L_p(\Omega)$$

(to be correct, here one needs the uniform cone condition on the domain, cf. 3.2), on the other, we have the following corollary :

$$E_{V_j}(f)_{H_p^s} \leq C 2^{-(s'-s)j} \|f\|_{H_p^{s'}} \, , \; f \in H_p^{s'} \, , \, 0 \leq s < 1 + 1/p \, , \; s < s' \leq 2 \, ,$$

$$\|g_j\|_{H_p^{s'}} \leq C 2^{(s'-s)j} \|g_j\|_{H_p^s} \, , \; g_j \in V_j \, , \; j \geq 0 \, , \, 0 \leq s < s' < 1 + 1/p \, .$$

For the direct estimate $q = 1$, and $q' = \infty$ has to be taken in the corresponding estimate of Theorem 11, for the Bernstein-type inequality $q = \infty$, $q' = 1$ will work. This result gives a complete picture of linear f.e. approximation theorems with respect to a sequence $\{T_j\}$ possessing the above regularity assumptions, in H_p^s spaces. It is easy to observe that the ranges for s, s' as well as the asymptotical order in the above estimates are optimal.

The reasoning can be extended to other f.e. schemes satisfying our above theory. The extension to slit domains is also possible and requires the same decomposition trick as used in 2.3, see also 3.4.5. For integer parameters s, s' direct and inverse estimates can be found in almost all books on finite elements, sometimes only for p=2, cf. [Ci, CO, Br1]. The proofs via Besov space techniques although restricted to special sequences of triangulations give the more general and final picture.

3.4.3 Norms on V_j and special representations

For applications, it is important to study norms on finite-dimensional spaces, e.g., on the V_j (or on subspaces occuring in adaptive processes (local refinement) which will be discussed in 4.2.2 resp. 5.3). We give a few examples.

First of all, due to Theorem 9, we may restrict ourselves to arbitrary finite representations in $(\mathbf{R})$ when $s > 0$, i.e. V_j equipped with the $A_{p,q}^s$ norm coincides with $A_{p,q}^s(\{V_l\}_{l \leq j})$, uniformly with respect to $j \geq 0$. Indeed, the necessary norm equivalence

$$\|g\|_{A^s_{p,q}} \approx \inf_{g_l \in V_l \,:\, g = \sum_{l=0}^{j} g_l} \|\{a^{sl} g_l\}_{l \le j}\|_{l_q(L_p)}$$

trivially follows in one direction by definition (finite representations are a special case of admissible representations). For the other direction requiring the construction of a good finite decomposition, one may use the best approximations g_l^*, $l \le j$ (due to $(\mathbf{M})$, we have $E_{V_l}(g)_p = 0$ and $g_l^* = g_j$, $l \ge j$!). The representation

$$g = g_0^* + \sum_{l=1}^{j} (g_j^* - g_{l-1}^*) \equiv \sum_{l \le j} g_l$$

will do the job (use the norm $\|\cdot\|_{A^s_{p,q}}^{+}$ from Theorem 9). Having this in mind, we will assume in the following that the $A^s_{p,q}$ norm of some finite element function $g \in V_j$ is defined by means of this restricted notion of finite admissible representations, i.e. we will use the same notation in a somewhat different (but equivalent) fashion. The other norms in Theorem 9 are either automatically defined by finite sums or will be modified in the same way.

We will give one more example of projections which produce some practically interesting decomposition, the sequence of natural interpolation projections I_j associated with the construction of the f.e. subspaces (another example are the quasi-interpolant projections $\mathcal{Q}_j$ introduced in subsection 2.1.1 which however have not yet found direct applications to multilevel schemes). Here, the details depend on the type of interpolation conditions used. Let

$$\|f\|_{I^s_{p,q}} = \|\{a^{sj}(I_j f - I_{j-1} f)\}\|_{l_q(L_p)} \qquad (I_{-1} f = 0) \, .$$

This quantity is formally defined ($< \infty$ or $= \infty$) if f belongs to the domain of all I_j. E.g., for C^0 Lagrange elements we can take $f \in C(\bar{\Omega})$. Anyhow, it is defined and finite for arbitrary $g \in V_j$, $j \ge 0$. Obviously, for these g and arbitrary parameters we have

$$\|g\|_{A^s_{p,q}} \le \|g\|_{I^s_{p,q}} \, .$$

For the other direction, since for any finite representation of g

$$I_l g - I_{l-1} g = g_l + I_l \Big(\sum_{k=l+1}^{j} g_k \Big) - I_{l-1} \Big(\sum_{k=l}^{j} g_k \Big) \equiv g_l + \tilde{g}_l - \tilde{g}_{l-1} \, ,$$

it suffices to estimate the quantity $\|\{a^{sl} \tilde{g}_l\}_{l < j}\|_{l_q(L_p)}$ or, in a first step, the L_p norms of arbitrary $\tilde{g}_l$, $l < j$. We will present the details for a f.e. scheme using function and derivative evaluations up to order r' as interpolation conditions for the somewhat simpler case of $1 \leq p = q < \infty$ (the general case can be found in [Os5]). Denoting by $\nu_{l,i}$ the functional corresponding to the basis function $N_{l,i}$ of V_l and involving the evaluation of some directional or mixed r-th derivative $(r \leq r')$ at some nodal point of $\mathcal{T}_l$ then we have by (**A1**)

$$\|\tilde{g}_l\|_p \approx a^{-ld/p} \|\{\nu_{l,i}(\sum_{k=l+1}^{j} g_k)\}\|_{l_p} .$$

Now we use the local (piecewise) polynomial nature of the underlying f.e. construction once again. For each $k > l$, to each $\nu_{l,i}$ there corresponds a subdomain $K_{k;l,i} \in \mathcal{T}_k$ and a polynomial $p \equiv p_{m';k;l,i} \in \mathbf{P}_{m'}$ coinciding with g_k on $K_{k;l,i}$ and such that

$$
\begin{aligned}
|\nu_{l,i}(g_k)| \;=\; |\nu_{l,i}(p)| \;&\leq\; Ca^{-rl}|p|_{W^r_\infty(K_{k;l,i})} \\
&\leq\; Ca^{r(k-l)}\|p\|_{L_\infty(K_{k;l,i})} \;\leq\; Ca^{(r+d/p)k-rl}\|g_k\|_{L_p(K_{k;l,i})}
\end{aligned}
$$

Since we have only a finite number of different r we run the further estimation for each group separately. If $r < s - d/p$ we choose an $\epsilon > 0$ such that $s - r - d/p - \epsilon > 0$. Then (using Hölder's inequality, and the fact that each $K_{k;l,i}$ may occur only finitely many times)

$$\sum_{l=0}^{j-1} a^{l(sp-d)} \sum_i^{(r)} |\sum_{k=l+1}^{j} \nu_{l,i}(g_k)|^p$$

$$\leq\; C \cdot \sum_{l=0}^{j-1} a^{l(sp-d)} \sum_i^{(r)} a^{-l(r+\epsilon)p} \sum_{k=l+1}^{j} a^{k((r+\epsilon)p+d)} \|g_k\|^p_{L_p(K_{k;l,i})}$$

$$\leq\; C \sum_{l=0}^{j-1} a^{l(s-r-\epsilon-d/p)p} \sum_{k=l+1}^{j} a^{k((r+\epsilon)p+d)} \|g_k\|^p_p$$

$$\leq\; C \sum_{k=1}^{j} a^{k((r+\epsilon)p+d)} \|g_k\|^p_p \sum_{l=0}^{k-1} a^{l(s-r-\epsilon-d/p)p}$$

$$\leq\; C \|\{a^{sk} g_k\}\|^p_{l_p(L_p)}$$

($\sum^{(r)}$ means summation with respect to only those i with interpolation condition of degree r). Analogous estimations have to be run for the cases $r = s - d/p$ resp. $r > s - d/p$. The correct application of the Hölder inequality gives the same estimate but with an additional factor j^p resp. $a^{j(r+d/p-s)p}$. Since the worst case appears if $r = r'$, after adding the estimates for all r and taking the infimum with respect to all finite representations of g, we finally arrive at

Theorem 12. For $1 \leq p < \infty$ and arbitrary $g \in V_j$, $j > 0$, we have

$$\|g\|_{A_{p,p}^s} \leq \|g\|_{I_{p,p}^s} \leq C\|g\|_{A_{p,p}^s} \begin{cases} 1 & \text{for } s - d/p > r' \\ j & \text{for } s - d/p = r' \\ a^{j(r'+d/p-s)} & \text{for } s - d/p < r' \end{cases} .$$

For subspaces of linear f.e., such results (even for general p and q) are given in [Os5], together with examples showing the asymptotical correctness of the additional factors if $s - d/p \leq r'(= 0)$. Cf. also [Os1, DOS]. The practically important case $s = 1, p = 2$ of Theorem 13 is due to Yserentant [Ys2] who proved this using a discrete Sobolev estimate, see [Df1] for generalizations along these lines. If $p = 2$ there is another, better candidate for the projection: the operator of best approximation $Q_j : L_2(\Omega) \to V_j$ is a linear projection, with $\|Q_j\|_{L_2 \to L_2} = 1$. Thus, Theorem 9 (cf. the above remarks) directly yields

$$\|g\|_{A_{2,2}^s}^2 \approx \|Q_0 g\|_2^2 + \sum_{l=1}^{j} \|Q_j g - Q_{j-1} g\|_2^2 \qquad \forall g \in V_j , \quad j \geq 0.$$

Note that there are many further modifications and problems which might be studied successfully within the context of sequences of finite-dimensional subspaces rather than in the infinite-dimensional space they approach (e.g. I_l is not well-defined on the whole $A_{p,p}^s(\{V_j\})$ if $s - d/p \leq r'$ since the embedding into $C^{r'}$ fails to hold but a useful theory can still be obtained by looking at the asymptotics of $A_{p,p}^s$ norms of I_l on V_j, $j \geq l$). However, the point is that the knowledge about the infinite-dimensional case still gives some important guidelines how to approach the finite-dimensional problems, and what is expected to happen there.

3.4.4 Extensions, traces, boundary conditions

We present a simple extension procedure for functions given on a polyhedral domain Ω onto $\mathbf{R}^d$ which preserves $A_{p,q}^s$ norms (and therefore also $B_{p,q}^{s;m}$ norms for the corresponding range of parameters). If Ω is polyhedral and satisfies the uniform cone condition (= is not a slit domain) then the partitions $\mathcal{T}_j$ corresponding to a f.e. scheme as described at the beginning of 3.4 can obviously be extended to the whole of $\mathbf{R}^d$, without destroying their regularity and quasi-uniformity substantially (the corresponding constants γ_i may change but stay independent of j). This is indicated in Figure 11 for the domain already shown in Figure 10. The new partitions $\tilde{\mathcal{T}}_j$ will produce f.e. subspaces $\tilde{V}_j$ defined on $\mathbf{R}^d$ the restriction of which to Ω coincides with V_j, $j \geq 0$. Clearly, **(A1)'** re-

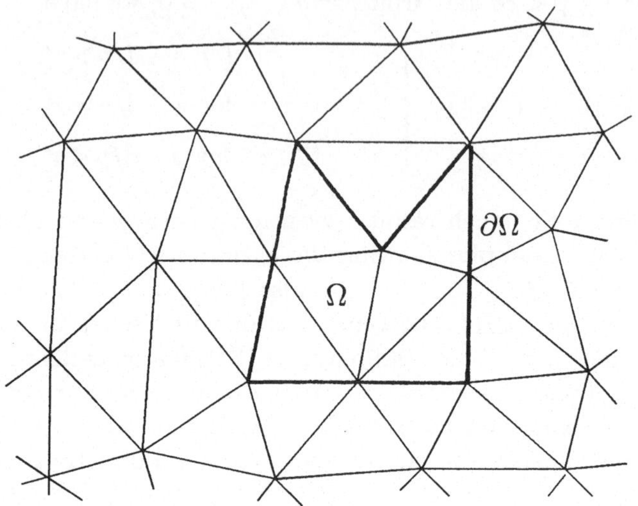

Figure 11. Extension of partitions $T_0 \to \tilde{T}_0$

mains true for these new subspaces, too. A few words about the new nodal basis $\{\tilde{N}_{j,i}\}$: by the construction of f.e. subspaces from local interpolation problems (which we take the same for both $\{T_j\}$ and $\{\tilde{T}_j\}$) it is clear that

$$\{N_{j,i}\} = \{\tilde{N}_{j,i'}|_\Omega \mid \operatorname{supp} \tilde{N}_{j,i'} \cap \Omega \neq \emptyset\}$$

For simplicity, let us enumerate the functions in the basis of $\tilde{V}_j$ so that $N_{j,i} = \tilde{N}_{j,i}|_\Omega$ for $i = 1, \ldots, n_j = \dim V_j$.

Take a (distinguished or anonymous) admissible representation (**R**) of $f \in A^s_{p,q}(\{V_j\})$ such that

$$\|\{a^{js}g_j\}\|_{l_q(L_p)} \leq C\|f\|_{A^s_{p,q}} \, ,$$

with some C a priori fixed. Each g_j possesses a basis representation

$$g_j = \sum_{i=1}^{n_j} c_{j,i} N_{j,i} \, ,$$

from which we produce new f.e. functions

$$\tilde{g}_j = \sum_{i=1}^{n_j} c_{j,i} \tilde{N}_{j,i} \in \tilde{V}_j \, .$$

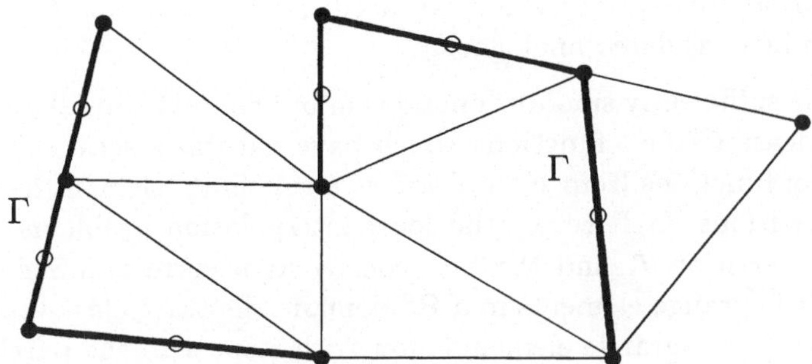

Figure 12. Admissible trace manifold Γ

with the obvious properties

$$\tilde{g}_j|_\Omega = g_j \quad , \qquad \|\tilde{g}_j\|_{L_p(\mathbf{R}^d)} \approx a^{-jd/p}\|\{c_{j,i}\}\|_{l_p} \approx \|g_j\|_p \quad \forall j .$$

Thus, we have $\|\{a^{sj}\tilde{g}_j\}\|_{l_q(L_p(\mathbf{R}^d))} < \infty$ which immediately yields the convergence (in $L_p(\mathbf{R}^d)$) of the series

$$\sum_{j=0}^{\infty} \tilde{g}_j = \tilde{f} \in L_p(\mathbf{R}^d) .$$

This is the desired extension.

If we do not specialize the representation $(\mathbf{R})$ this extension is not a linear mapping and depends also from the different parameters s, p, q. We can take the quasi-interpolants from 2.1.1 to construct a decomposition: this results in a linear extension operator $\mathcal{E}$ which works for all $A_{p,q}^s$ spaces associated with the fixed f.e. approximation scheme $\{V_j\}$ simultaneously!

Now we turn to the trace problem which will be studied in a restricted but still sufficient general fashion. Let Γ denote any regular $\bar{d}$-dimensional part of the boundary of $\mathcal{T}_0$, i.e. Γ is the union of $\bar{d}$-dimensional faces of subdomains in $\mathcal{T}_0$, see Figure 12. Prominent example, with $\bar{d} = d - 1$, is the boundary $\partial\Omega$ of the domain. Below we will also discuss a further example where Γ is the whole boundary of some $\mathcal{T}_{j_0}$ which is of interest in some variant of the domain decomposition method. In accordance with the general definition of trace spaces mentioned in 3.3, we are going to describe the trace spaces for $A_{p,q}^s(\{V_j\})$ which are, roughly speaking, the closures of the set of ordinary traces $h = f|_\Gamma$ of sufficiently smooth functions $f \in A_{p,q}^s(\{V_j\})$ with respect to the norm

$$\|h\|_{A_{p,q}^s|_\Gamma} := \inf_{f \in A_{p,q}^s : h = f|_\Gamma} \|f\|_{A_{p,q}^s} .$$

Traces of various derivatives which we need, for instance, for the biharmonic problem can be considered analogously.

What are the sufficiently smooth functions in our context? In all our examples, we have at least C^0-f.e. functions which have natural traces. Moreover, the restrictions of functions from V_j to Γ are as a rule finite element like functions, with the partitions on Γ resp. the local interpolation problems induced by the original partition $\mathcal{T}_j$ and the f.e. construction corresponding to V_j. For simplicial C^0 Lagrange elements in a $\mathbf{R}^d$-domain, the trace class $V_j|_\Gamma$ is nothing else but the same Lagrange elements, now with respect to the partition $\mathcal{T}_j|_\Gamma$ of a $\bar{d}$-dimensional piecewise simplicial manifold Γ. Details have to be checked for the particular f.e. scheme. What we really need is summarized in the following property.

(A1-Γ) Suppose that the set of all nontrivial traces $N_{j,i}|_\Gamma \not\equiv 0$ of nodal basis functions to Γ forms a $L_p(\Gamma)$-stable basis of $V_{j,\Gamma}$. More precisely, let

$$\|h_j\|_{L_p(\Gamma)} \approx a^{j(d-\bar{d})} \inf_{g_j \in V_j\,:\,h_j=g_j|_\Gamma} \|g_j\|_p \quad \forall\, h_j \in V_j|_\Gamma\,,\ 1 \le p \le \infty\,.$$

This is obvious for C^0 Lagrange elements, for C^1 elements and derivatives it has to be checked correspondingly.

With this property at hand, the proof of the following theorem can be given within a few lines. It shows that trace spaces can alson be characterized in terms of $A^s_{p,q}$-spaces.

Theorem 13.   Let the sequence of finite element spaces satisfy in addition the property **(A1-Γ)**. Then for $s > (d-\bar{d})/p$

$$A^s_{p,q}(\{V_j\})|_\Gamma = A^{s-(d-\bar{d})/p}_{p,q}(\{V_j|_\Gamma\})$$

with equivalent norms (the second space is defined with respect to $L_p(\Gamma)$).

Proof. First we show that any $f \in A^s_{p,q}(\{V_j\})$ possesses a meaningful trace $h = f|_\Gamma \in L_p(\Gamma)$. Consider any admissible representation **(R)** and denote $h_j = g_j|_\Gamma$. The series $\sum_{j=0}^\infty h_j$ converges in $L_p(\Gamma)$ since by **(A1-Γ)**

$$\sum_{j=0}^\infty \|h_j\|_{L_p(\Gamma)} \le C\|\{a^{(s-(d-\bar{d})/p)j} h_j\}\|_{l_q(L_p(\Gamma))} \le C\|\{a^{sj} g_j\}\|_{l_q(L_p)} < \infty\,.$$

Thus, $h = \sum_{j=0}^\infty h_j$ is well-defined. In addition, the estimation shows that $h \in$

$A_{p,q}^{s-(d-\bar{d})/p}(\{V_j|_\Gamma\})$ with

$$\|h\|_{A_{p,q}^{s-(d-\bar{d})/p}(\Gamma)} \leq C\|f\|_{A_{p,q}}$$

It remains to prove that h is indeed the trace of f (it does not depend on the choice of the admissible representation in $(\mathbf{R})$, and continuously depends on f in the respective norms). A simple way to do this is to choose in the above construction not a general representation of f but the quasi-interpolant decomposition which obviously produces a linear bounded trace operator tr_Γ mapping $f \in A_{p,q}^s(\{V_j\})$ to $h = tr_\Gamma(f) \in A_{p,q}^{s-(d-\bar{d})/p}(\{V_j|_\Gamma\})$. To do this, we need assumption $(\mathbf{A2})$. A detailed proof not using this argument is given in [Os1].

Inversely, let $h \in A_{p,q}^{s-(d-\bar{d})/p}(\{V_j|_\Gamma\})$ arbitrarily be given. Consider, for this $A_{p,q}^s$-space with respect to Γ, any admissible representation of h as a sum of $h_j \in V_j|_\Gamma$. By $(\mathbf{A1}\text{-}\Gamma)$ there exist such $g_j \in V_j$ that $h_j = tr_\Gamma(g_j)$ and

$$\|g_j\|_p \leq Ca^{-j(d-\bar{d})/p}\|h_j\|_{L_p(\Gamma)} \qquad \forall j \geq 0 \ .$$

As above, this yields the L_p convergence of $\sum_{j=0}^\infty g_j$ to some $f \in A_{p,q}^s(\{V_j\})$, with

$$\|f\|_{A_{p,q}^s} \leq C\|h\|_{A_{p,q}^{s-(d-\bar{d})/p}(\Gamma)}.$$

To finish the argument, one has to check whether $tr_\Gamma(f) = h$. But this is obvious from the first part: The trace operator tr_Γ acts continuously between $A_{p,q}^s(\{V_j\})$ and $A_{p,q}^{s-(d-\bar{d})/p}(\{V_j|_\Gamma\})$ whereas

$$\sum_{j=0}^n g_j \to f \ , \qquad \sum_{j=0}^n h_j = tr_\Gamma\left(\sum_{j=0}^n g_j\right) \to h$$

in the respective A-norms. Theorem 13 is established.

Note that traces of higher-order derivatives (which are meaningful in the present context only if the f.e. scheme allows for them!) can be dealt with in analogy. Of interest is, e.g., the mapping $f \to \frac{\partial f}{\partial n}|_\Gamma$, both for C^1 elements where it can be handled as before [Os8], and for C^0 elements in connection with Dirichlet-Neumann mappings etc.. In the latter case we see the difficulty - our $A_{p,q}^s$ spaces are restricted to positive smoothness parameters, and a theory of dual spaces has not yet been developed. We come back to this point later on.

We do not make any attempt to find characterizations of the A-spaces defined on Γ in terms of derivatives, differences etc., it turns out that for many numerical applications such knowledge is not really necessary. Clearly, one can do this in the same way as in section 2 developing the Jackson-Bernstein machinery (or other techniques, see 3.1.2) now with respect to f.e. subspaces on piecewise polyhedral manifolds. Moreover, the results are known in some cases: [Gv] contains a discussion of trace spaces for Sobolev spaces in polygonal domains.

With Theorem 13 at hand, one can now define spaces that are of interest for boundary value problems. Once again, we treat only the case of second order elliptic equations where the only essential boundary conditions are Dirichlet conditions. We restrict our construction to the case that $\Gamma(\subset \partial\Omega)$ is the union of closed $d-1$-dimensional faces of subdomains of $\mathcal{T}_0$, i.e. we suppose that $\mathcal{T}$ resolves $\partial\Omega$ such that the Dirichlet part (represented by Γ) is exactly covered by a certain number of subdomains. There are two possibilities of defining the $A^s_{p,q}$-space of functions that vanish at Γ : as above we can take the subset of all contionuous (or C^∞) functions $f \in A^s_{p,q}(\{V_j\})$ with $f(x) = 0$ on Γ and look at its closure, or, in view of Theorem 13, one could define

$$A^s_{p,q,D=\Gamma}(\{V_j\}) = \{f \in A^s_{p,q}(\{V_j\}) \mid tr_\Gamma(f) = 0\}, \quad s > 1/p .$$

It is almost clear that there is no difference between these two definitions if $s > 1/p$.

Theorem 14. Under the assumptions of Theorem 13 ($\bar{d} = d - 1$), for $s > 1/p$, we have

$$A^s_{p,q,D=\Gamma}(\{V_j\}) = A^s_{p,q}(\{V_{j,D=\Gamma}\}), \quad V_{j,D=\Gamma} = \{g_j \in V_j \mid tr_\Gamma(g_j) = 0\}$$

with equivalent norms.

We omit the proof. Note that the space in the right-hand side makes sense also for $0 \leq s \leq 1/p$. It can be shown that for $s < 1/p$ it coincides with $A^s_{p,q}(\{V_j\})$, i.e. there is no chance of controlling boundary behaviour in classes with low smoothness - a well-known fact.

Once again we want to refer to [LM] and especially to [Gv] concerning the corresponding discussions in terms of usual Besov-Sobolev norms. Our results are very particular, but simple and closely adapted to the problems we want to attack in section 4 : fast multilevel methods for finite element discretizations.

3.4.5   Output: Decomposition norms in H^s

This short subsection is a service for those who do not like many indices and L_p theory with $p \neq 1, 2, \infty$ (we come back to $p \neq 2$ in section 5). We subsume the results for $p = q = 2$, for an f.e. scheme as described at the beginning of 3.4 but with $a = 2$ (regular dyadic refinement) and $r < m - 1$ which also satisfies **(A2)**, this guarantees the Jackson-Bernstein theory of section 2 and allows us to switch from $A_{2,2}^s$ to $H^s = B_{2,2}^{s,m}$ for the corresponding values of s and m. The domain $\Omega \subset \mathbf{R}^d$ will be polyhedral, bounded, and satisfy the uniform cone condition. Slit domains for which the theorems below hold true are discussed in a remark below. In the case of spaces with essential boundary conditions, we assume $(\mathbf{A1}\text{-}\Gamma)$ for $\bar{d} = d - 1$ where Γ is such as required for Theorem 14, and restrict ourselves to Dirichlet conditions involving only $tr_\Gamma(f)$ (see 4.3 where some results for C^1 elements and traces of derivatives are discussed and applied).

Theorem 15. Under the above assumptions, the following assertions hold true (with constants depending on s, the finite element type, especially on r, m, and m') and the geometry of the initial partition $\mathcal{T}_0$)

- For all $f \in H^s(\Omega)$ and $0 < s < r + 3/2$

$$\|f\|_{H^s}^2 \approx \inf_{g_j \in V_j \,|\, f = \sum_j g_j} \sum_{j=0}^{\infty} 2^{2sj} \|g_j\|_2^2$$

- For all $g_j \in V_j$, $j \geq 0$, and $0 < s < r + 1 + 1/p$

$$\|g_j\|_{H^s}^2 \approx \inf_{g_l \in V_l \,|\, f = \sum_{l=0}^{j} g_l} \sum_{l=0}^{j} 2^{2sl} \|g_l\|_2^2$$

$$\approx \inf_{g_l \in V_l \,|\, f = \sum_{l=0}^{j} g_l} \sum_{l=0}^{j} 2^{2(s-d/2)l} \sum_i c_{l,i}^2 \qquad \left(g_l = \sum_i c_{l,i} N_{l,i}\right)$$

$$\approx \|Q_0 g_j\|_2^2 + \sum_{l=1}^{j} 2^{2sl} \|Q_l g_j - Q_{l-1} g_j\|_2^2$$

where $Q_l : L_2(\Omega) \to V_l$ are the L_2-orthoprojections.

- For all $f \in H^{s'}(\Omega)$, and $0 \leq s < r + 3/2 \; s < s' \leq m$

$$E_{V_j}(f)_{H^s} \leq C 2^{-j(s'-s)} \|f\|_{H^{s'}}$$

(Jackson-type estimate - approximation property), and for all $g_j \in V_j$, $j \geq 0$, and $0 \leq s < s' < r + 3/2$

$$\|g_j\|_{H^{s'}} \leq C 2^{j(s'-s)} \|g_j\|_{H^s}$$

(inverse property).

- The above assertions hold for $f \in H_{D=\Gamma}^s(\Omega)$ resp. $g_j \in V_{j,D=\Gamma}$ if V_j, V_l are replaced by $V_{j,D=\Gamma}, V_{l,D=\Gamma}$ and s (resp. s') is restricted to $1/2 < s < r + 3/2$.

In addition, we could rewrite Theorem 12 (comparison of the decomposition norm with respect to the natural interpolation projections with the H^s norms) and Theorem 13 (description of $H^s(\Gamma)$, at least for $0 < s < 1$). We leave this up to the reader. Note that in [BY] the practically most important case of linear finite elements, $s = 1$, of the above theorem has been examined, slightly different proofs are also available from other authors (see 4.2 for more details).

Remark. Are there any problems with slit domains? We have already seen (subsection 2.3) that a domain splitting argument may be explored to overcome the problems. Let us assume that we have a covering of T_0 resp. Ω by a finite number of subpartitions $T_0^i \subset T_0$ resp. corresponding open domains Ω^i, $i = 1, \ldots, n$, such that T_0^i is an initial partition of Ω^i, and the Ω^i satisfy the uniform cone condition. Covering implies, e.g., that $\Omega = \cup \Omega^i$ which by a simple induction argument implies that (after a possible reordering of the subdomains) $\Omega^{i+1} \cap \cup_{k \leq i} \Omega^i$ contains at least one subdomain K^i of T_0. Thus, we assume a decomposition of Ω into nice domains compatible with T_0 and with enough overlap. Clearly, to each Ω^i there corresponds a sequence $V_j^i = V(T_j^i) = V_j|_{\Omega^i}$ which produces scales $A_{p,q}^{s,i} \equiv A_{p,q}^s(\{V_j^i\})$. Denote $f^i = f|_{\Omega^i}$.

A close look at the definitions of B- and A-spaces and at the proof of the norm equivalence in Theorem 6 shows the inequalities

$$\|f\|_{B_{p,q}^{s,m}} \leq C\|f\|_{A_{p,q}^s} \leq C \sum_i \|f^i\|_{A_{p,q}^{s,i}} \leq C \sum_i \|f^i\|_{B_{p,q}^{s,m}(\Omega^i)} \leq C\|f\|_{B_{p,q}^{s,m}}$$

for the range of parameters indicated in Theorem 6. This is the proof that for the norm equivalence between B- and A-spaces the uniform cone condition is superfluous. The point where we have to refer to the remark after Theorem 2 in 2.3, is the second inequality: from suitable representations for each f^i one has to construct an equally good one for f itself. Here the overlap is used. Spaces with conditions on the trace can be dealt with by the same trick.

Now we want to have the analogous argument for the comparison with H^s instead of $B_{2,2}^{s,m}$ (as an open problem we may ask for a direct proof of $H^s(\Omega) = B_{2,2}^{s,m}(\Omega)$, $0 < s < m$, without using extension arguments, at least for integer s and arbitrary polygonal domains). What we can rely on is the coincidence of the two spaces on each of the Ω_i (see Theorem 7 and 8). For integer $s = k$ (within the range of admissible parameters) we have

$$\|f\|_{H^k(\Omega)}^2 \approx \sum_i \|f^i\|_{H^k(\Omega^i)}^2$$

which in more precise terms means that the space of all $f \in L_2(\Omega)$ such that $f^i \in H^k(\Omega^i)$ for $i = 1, \ldots, n$ coincides with $H^k(\Omega)$ plus the above norm equivalence.

This is trivial for each finite open covering of an open domain (the constants depend on n which does not matter in the present context). Thus, we can complete our argument

$$\|f\|^2_{H^k(\Omega)} \approx \sum_i \|f^i\|^2_{H^k(\Omega^i)} \approx \sum_i \|f^i\|^2_{B^{k,m}_{2,2}(\Omega^i)} \approx \|f\|^2_{B^{k,m}_{2,2}(\Omega)} \approx \|f\|^2_{A^k_{2,2}(\Omega)}$$

for all $k = 1, \ldots, r+1$. The same result holds true for $H^k_{D=\Gamma}$. Unfortunately, we are not sure whether the decomposition argument is valid for arbitrary $s > 0$, too.

4 Applications to multilevel methods

We deal with the basic theory of finite element multilevel methods for symmetric positive definite variational problems in a Hilbert space. In subsection 4.1 we introduce the so-called additive and multiplicative Schwarz algorithms based on appropriate subspace splittings which include multilevel methods as particular case. The convergence theory for this class of iterative schemes has a long history but has been put into a more or less final, abstract form only recently. We rely on the surveys by Xu [Xu1], Yserentant [Ys1], and on [GO2] but are influenced also by papers of Dryja/Widlund [DW1, Wi], Nepomnyaschikh [Ne1, Ne2], and Zhang [Zh1]. The remaining sections discuss special cases of computationally relevant splittings of Sobolev spaces into multilevel finite element spaces. We give sharp estimates for the corresponding algorithms (multilevel preconditioners, domain decomposition methods etc.) which reproduce or improve recent results of many authors, or are new. Our approach illustrates the usefulness of the norm equivalences from approximation theory developed in the previous sections.

4.1 The abstract Schwarz theory

We use the following notation. Let V be some fixed separable Hilbert space (at the moment we do not assume that V is finite-dimensional). The scalar product in V is denoted by $(\cdot, \cdot)_V$. We consider a symmetric positive definite (s.p.d.) bilinear form $a(u, v) = (Au, v)_V$, $u, v \in V$, with $A : V \to V$ denoting the corresponding s.p.d. operator acting on V (for the purposes of our exposition, we will identify the dual V' with V via the Riesz-Fischer theorem). Below we often take $a(\cdot, \cdot)$ as the basic scalar product on V which is justified by the above requirements and will be indicated by writing $\{V; a\}$ instead of simply V. The spectral condition number of a s.p.d. operator P in $\{V; a\}$ is defined by

$$\mathrm{cond}\,(P) = \frac{\lambda_{\max}(P)}{\lambda_{\min}(P)} \,.$$

If V is finite-dimensional then the numbers $\lambda_{\min}(P)$ resp. $\lambda_{\max}(P)$ coincide with the smallest resp. largest eigenvalues of P. Generally, they are defined by

$$\lambda_{\min}(P) = \inf_{u \in V,\, u \neq 0} \frac{a(Pu, u)}{a(u, u)} \;, \quad \lambda_{\max}(P) = \sup_{u \in V,\, u \neq 0} \frac{a(Pu, u)}{a(u, u)} \;.$$

We look for solution procedures for a variational problem associated with the bilinear form $a(\cdot, \cdot)$: Find $u \in V$ such that

(VP) $a(u, v) = \Phi(v)$, $\forall\, v \in V$.

Here $\Phi(v) = (f, v)_V$ is an arbitrary linear functional on V ($f \in V$ is given). In operator form, **(VP)** is equivalent to $Au = f$. Due to the Lax-Milgram theory, **(VP)** possesses a unique solution.

The iterative solvers for this problem we are looking for are based on the decomposition of V into a (not necessarily direct) sum of subspaces $V_j \subset V$:

$$V = \sum_j V_j \;,$$

i.e., we assume that any $u \in V$ possesses at least one V-converging decomposition

$$u = \sum_j u_j \;, \qquad u_j \in V_j \;\; \forall\, j \;.$$

The number of subspaces may be finite or infinite. However, if we come to algorithms, we will suppose that $j = 0, \ldots, J$, and that all Hilbert spaces are finite-dimensional. On the subspaces we choose auxiliary s.p.d. forms $b_j(u_j, v_j) = (B_j u_j, v_j)_V$ given by the s.p.d. operators $B_j : V_j \to V_j$. These forms model the approximate solvers used in the subspaces, i.e. B_j^{-1} is an approximate inverse for the restriction A_j of A to V_j. Furthermore, let $T_j : V \to V_j$ be given by the variational problems

$$b_j(T_j u, v_j) = a(u, v_j) \;, \qquad \forall\, v_j \;, u \in V \;, j = 0, \ldots, J$$

(in operator notation $T_j = B_j^{-1} Q_j A$ where $Q_j : V \to V_j$ denotes the orthoprojection onto V_j with respect to the scalar product $(\cdot, \cdot)_V$). Finally, let $\phi_j \in V_j$ be the solution of the problems

$$b_j(\phi_j, v_j) = \Phi(v_j) \;, \qquad \forall\, v_j \;, j = 0, \ldots, J \;.$$

Before we formulate the first basic result, we introduce the notion of a *stable subspace splitting* which is important for the understanding of the convergence theory of the Schwarz methods. Since the definition involves not only the above decomposition into a sum of subspaces but also the s.p.d. bilinear forms $a(\cdot, \cdot)$

resp. $b_j(\cdot,\cdot)$ given on V resp. on V_j, we will rewrite the subspace splitting in the following, more explicit way:

$$(\mathbf{D}) \qquad \{V;a\} = \sum_j \{V_j;b_j\} \; .$$

The splitting $(\mathbf{D})$, with the bilinear forms fixed as indicated, is called *stable* if a two-sided inequality

$$(\mathbf{NE}) \quad a(u,u) \approx |||u|||^2 \equiv \inf_{u_j \in V_j \, : \, u = \sum_j u_j} \sum_j b_j(u_j,u_j) \qquad \forall \, u \in V$$

holds true. The best constants in the norm equivalence $(\mathbf{NE})$ will be denoted by

$$0 < \lambda_{\min} = \min_{u \in V, \, u \neq 0} \frac{a(u,u)}{|||u|||^2} \leq \lambda_{\max} = \max_{u \in V, \, u \neq 0} \frac{a(u,u)}{|||u|||^2} < \infty \;\; ,$$

and the number $\kappa = \lambda_{\min}/\lambda_{\max}$ will be called stability constant or *condition number* of the splitting $(\mathbf{D})$. Note that for finite splittings of a finite-dimensional V, stability is automatically fulfilled and the question which remains is to estimate the size of the positive constants $\lambda_{\min}, \lambda_{\max}$, and κ. The role which these quantities play for subspace correction methods will become clear very soon.

For infinite splittings, the stability implies the unconditional convergence of the series $\sum_j T_j u$ (for arbitrary $u \in V$) resp. $\sum_j \phi_j$ in the Hilbert space V. Without loss of generality, it is sufficient to prove this for the case $V_0, V_1, \ldots$. Denote $v_j = T_j u$ and fix an arbitrary integer J. Then we have for $v = \sum_{j=0}^{J} v_j$ by the above definitions

$$\lambda_{\max}^{-1} a(v,v) \leq \sum_{j=0}^{J} b_j(v_j,v_j) = \sum_{j=0}^{J} a(u,v_j) = a(u,v) \leq \sqrt{a(u,u)}\sqrt{a(v,v)}$$

which gives $a(v,v) \leq \lambda_{\max}^2 a(u,u)$ and

$$\sum_{j=0}^{J} b_j(T_j u, T_j u) \leq \lambda_{\max} a(u,u) \qquad \forall J \quad \forall \, u \in V \; .$$

Thus, the series $\sum_{j=0}^{\infty} b_j(T_j u, T_j u)$ converges for each individual $u \in V$, and by the Cauchy criteria

$$\| \sum_{j=m}^{n} T_j u \|_V^2 \approx a(\sum_{j=m}^{n} T_j u, \sum_{j=m}^{n} T_j u) \leq \lambda_{\max} \sum_{j=m}^{n} b_j(T_j u, T_j u) \to 0 \; ,$$

for $m, n \to \infty$ which implies the convergence of $\sum_j T_j u$. As a by-product, we see that the operator

$$P = \sum_j T_j \; : \; V \to V \;,$$

is linear and bounded on V. The completely analogous proof of the convergence of $\phi = \sum_j \phi_j$ is left up to the reader.

Theorem 16. (Additive Schwarz preconditioner) Let the splitting **(D)** be stable. Then, the solution of **(VP)** coincides with that of the operator equation

(AS) $Pu = \phi$

where the additive Schwarz operator $P = \sum_j T_j$ associated with the splitting **(D)** is s.p.d. in $\{V; a\}$, and $\phi = \sum_j \phi_j$. Moreover,

$$\lambda_{\min}(P) = \lambda_{\min} \;, \quad \lambda_{\max}(P) = \lambda_{\max} \;, \quad \mathrm{cond}(P) = \kappa \;,$$

with the characteristic constants of the splitting defined after **(NE)**.

This theorem is very simple but useful: it says under which necessary and sufficient conditions on the splitting **(D)** and the choice for the forms b_j the new problem **(AS)** is well-conditioned. The idea is then to solve **(AS)** by Richardson or cg-iterations (cf. [AB, Ha2]) which should lead to fast convergence due to the small condition number of the additive Schwarz operator. Since

$$P = \sum_j T_j = (\sum_j B_j^{-1} Q_j) A \equiv CA \;\;,$$

the procedure can be viewed as preconditioning strategy with preconditioner C for the original problem.

Theorem 16 has many authors, and it is not our intention to find out all contributors (as with any simple thing, people go around for many years), we learned about it from the papers by Widlund et.al. (see [Wi, BM, Zh1]). We give a derivation via the following fictitious space lemma proved by Nepomnyaschikh (see [Ne1, Ne2]) some years ago which is useful on its own for different purposes.

Theorem 17. (Fictitious space lemma). Let V and $\tilde{V}$ be two Hilbert spaces, with the scalar products denoted by $(\cdot,\cdot)_V$ resp. $(\cdot,\cdot)_{\tilde{V}}$, and with bilinear forms $a : V \times V \to \mathbf{R}$ resp. $\tilde{a} : \tilde{V} \times \tilde{V} \to \mathbf{R}$ generated by the s.p.d. operators $A : V \to V$ resp. $\tilde{A} : \tilde{V} \to \tilde{V}$. Suppose that there is a surjective bounded linear operator $R : \tilde{V} \to V$ such that the following two-sided inequality is satisfied:

$$a(u,u) \approx |||u|||_*^2 \equiv \inf_{\tilde{v} \in \tilde{V} \,:\, u = R\tilde{v}} \tilde{a}(\tilde{v}, \tilde{v}) \qquad \forall\, u \in V \;.$$

Then the operator $P = R\tilde{A}^{-1}R^*A : v \to V$ is s.p.d. in $\{V; a\}$, with

$$\lambda_{\min}(P) = \inf_{u \in V,\, u \neq 0} \frac{a(u,u)}{|||u|||_*^2}\,, \qquad \lambda_{\max}(P) = \sup_{u \in V,\, u \neq 0} \frac{a(u,u)}{|||u|||_*^2}\,.$$

The adjoint operator $R^* : V \to \tilde{V}$ is given by $(R\tilde{v}, v)_V = (\tilde{v}, R^*v)_{\tilde{V}}$ for all $v \in V$ and $\tilde{v} \in \tilde{V}$.

Proof (cf. [Ne2]). P is symmetric since

$$\begin{aligned}
a(Pu,v) &= (R\tilde{A}^{-1}R^*Au, Av)_V = (\tilde{A}^{-1}R^*Au, R^*Av)_{\tilde{V}} \\
&= \tilde{a}(\tilde{A}^{-1}R^*Au, \tilde{A}^{-1}R^*Av)\,.
\end{aligned}$$

Setting $v = u$ we see that P is positive semi-definite, and that P is injective if R^* does so. But this follows from the surjectivity of R. Now, for any $\tilde{v} \in \tilde{V}$ such that $R\tilde{v} = Pu$ we have

$$\begin{aligned}
a(Pu,u) &= a(R\tilde{v},u) = (R\tilde{v}, Au)_V = (\tilde{v}, R^*Au)_{\tilde{V}} = \tilde{a}(\tilde{v}, \tilde{A}^{-1}R^*Au) \\
&\leq \sqrt{\tilde{a}(\tilde{v},\tilde{v})}\sqrt{\tilde{a}(\tilde{A}^{-1}R^*Au, \tilde{A}^{-1}R^*Au)} = \sqrt{\tilde{a}(\tilde{v},\tilde{v})}\sqrt{a(Pu,u)}\,,
\end{aligned}$$

with equality for $\tilde{v} = \tilde{A}^{-1}R^*Au$. Thus, taking the infimum with respect to $\tilde{v}$, we have the equality

$$a(Pu,u) = |||Pu|||_*^2\,, \quad u \in V\,.$$

In full analogy, considering arbitrary $\tilde{v} \in \tilde{V}$ with $R\tilde{v} = u$, we get

$$a(u,u) \leq |||u|||_* |||Pu|||_* \leq c\sqrt{a(u,u)}\sqrt{a(Pu,Pu)}\,,$$

with a certain constant c which comes from applying the assumed norm equivalence. This shows the positive definiteness and invertibility of P.

Finally, putting $u = P^{-1}v$ in the above equality we obtain $a(P^{-1}v,v) = |||v|||_*^2$. Due to $\lambda_{\max}(P) = \lambda_{\min}(P^{-1})^{-1}$ and $\lambda_{\min}(P) = \lambda_{\max}(P^{-1})^{-1}$, this yields the desired result.

To see that Theorem 16 is a consequence of Theorem 17, we define

$$\tilde{V} = \{\tilde{v} = \{v_j\}_j : v_j \in V_j\,, \sum_j b_j(v_j, v_j)_V < \infty\}\,,$$

as the cartesian product of the Hilbert spaces $\{V_j; b_j\}$ which is obviously a Hilbert space with the bilinear form

$$(\tilde{u}, \tilde{v})_{\tilde{V}} \equiv \tilde{a}(\tilde{u}, \tilde{v}) = \sum_j b_j(u_j, v_j)$$

as the scalar product. Concerning V, it will be convenient to identify the scalar product $(\cdot,\cdot)_V$ with $a(\cdot,\cdot)$, i.e., we take $V = \{V; a\}$. Consequently, $\tilde{A}$ and A are the identity operators on the respective Hilbert spaces. As we did for the additive Schwarz operator P, by using the stability of (D), we can define a bounded linear operator $R : \tilde{V} \to V$ by the formula $R\tilde{v} = \sum_j v_j$. The surjectivity of R follows from the lower bound in (NE), i.e., from $\lambda_{\min} > 0$. Moreover, the two-sided estimate assumed in Theorem 17 coincides exactly with (NE). Since

$$a(u, \sum_j v_j) = (u, R\tilde{v})_V = (R^*u, \tilde{v})_{\tilde{v}} = \sum_j b_j((R^*u)_j, v_j)$$

for all $\tilde{v} \in \tilde{V}$ and $u \in V$ we obtain $R^*u = \{T_j u\}_j$ and, eventually, $R\tilde{A}^{-1}R^*A = RR^* = \sum_j T_j$. Thus, the additive Schwarz operator is s.p.d. in $\{V; a\}$, with the corresponding formulae for $\lambda_{\min}(P)$ and $\lambda_{\max}(P)$, and (AS) possesses a unique solution. That this solution coincides with that for (VP) is obvious by definition of $T_j u$ and ϕ_j. Theorem 17 is completely proved.

The above introduction of the Hilbert space $\tilde{V}$ as the cartesian product of the subspaces $\{V_j; b_j\}$ is a useful technical device. E.g., defining $\tilde{P} = R^*R : \tilde{V} \to \tilde{V}$ and $\tilde{\phi} = \{\phi_j\} \in \tilde{V}$, we can consider another operator equation

(A͠S) $\tilde{P}\tilde{u} = \tilde{\phi}$,

which is equivalent to (VP) in the following sense: if $\tilde{u} \in \tilde{V}$ is any solution of (A͠S) then $u = R\tilde{u}$ gives the unique solution of (VP). This follows from applying R to both sides and comparing with (AS). One can look at $\tilde{P}$ as at an operator matrix, with the entries

$$P_{ij} = T_i|_{V_j} : V_i \to V_j .$$

Since

$$b_i(P_{ij}u_j, v_i) = b_i(T_i u_j, v_i) = a(u_j, v_i) = b_j(u_j, T_j v_i) = b_j(u_j, P_{ji}v_i)$$

for all i, j, the operator $\tilde{P}$ is symmetric in $\tilde{V}$. Moreover,

$$\tilde{a}(\tilde{P}\tilde{u}, \tilde{u}) = \sum_i b_i(T_i(\sum_j u_j), u_i) = \sum_i a(R\tilde{u}, u_i) = a(R\tilde{u}, R\tilde{u}) \geq 0$$

shows the positive semi-definiteness, and the property $\operatorname{Ker}(\tilde{P}) = \operatorname{Ker}(R)$, i.e. $\tilde{P}$ has a nontrivial nullspace iff the splitting (D) is not a direct sum of subspaces.

We come to the Schwarz methods associated with a splitting (D) for iteratively solving the variational problem (VP). From now on, let the Hilbert space V be finite-dimensional, and the splitting (D) be into a finite number of $J + 1$ subspaces V_j, $j = 0, \ldots, J$. As was mentioned before, under these assumptions

(D) is automatically stable and Theorem 16 holds true. We introduce a so-called *additive* and two slightly different *multiplicative* algorithms based on subspace corrections with respect to **(D)**.

Additive Schwarz method. Starting with an initial guess $u^{(0)} \in V$, repeat

$$u^{(k+1)} = u^{(k)} - \omega \sum_{j=0}^{J} (T_j u^{(k)} - \phi_j) , \ k \geq 0 ,$$

until a stopping criteria is satisfied.

Multiplicative Schwarz method. Starting with an initial guess $u^{(0)} \in V$, repeat for $k \geq 0$

$$v^{(0)} = u^{(k)}$$

$$v^{(j+1)} = v^{(j)} - \omega \sum_{j=0}^{J} (T_j v^{(j)} - \phi_j) , \quad j = 0, \ldots, J$$

$$u^{(k+1)} = v^{(J+1)}$$

until a stopping criteria is satisfied.

Symmetric multiplicative Schwarz method. Starting with an initial guess $u^{(0)} \in V$, repeat for $k \geq 0$

$$v^{(0)} = u^{(k)}$$

$$v^{(j+1)} = v^{(j)} - \omega (T_j v^{(j)} - \phi_j) , \quad j = 0, \ldots, J$$

$$v^{(2J+2-j)} = v^{(2J+1-j)} - \omega (T_j v^{(2J+1-j)} - \phi_j) , \quad j = J, \ldots, 0$$

$$u^{(k+1)} = v^{(2J+2)}$$

until a stopping criteria is satisfied.

What is common for all three variants is that per iteration step $J+1$ (or twice as many) subproblems of the form

$$b_j(r_j, v_j) = a(u, v_j) - \Phi(v_j) \quad \forall \, v_j \in V_j ,$$

with some given $u \in V$ have to be solved. These determine the updates $r_j = T_j u - \phi_j \in V_j$ entering the iteration. Note that the additive method looks more parallel since the $J+1$ subproblems to be solved within one iteration do not interact.

To see that these methods are the abstract counterparts of the classical Jacobi-Richardson, SOR, and SSOR methods for linear systems in $\mathbf{R}^n$, resp., we rewrite them using the $(J+1) \times (J+1)$ operator matrix $\tilde{P}$. First we decompose

$$\tilde{P} = \tilde{L} + \tilde{D} + \tilde{U}$$

into strictly lower triangular, diagonal, and upper triangular parts. From above we already know that $\tilde{U} = \tilde{L}^*$ (in the sense of operators in $\{\tilde{V}, \tilde{a}\}$). Let I and $\tilde{I}$ denote the identity operator on V and $\tilde{V}$, resp.. Consider a linear iteration method of the form

$$\tilde{u}^{(k+1)} = \tilde{u}^{(k)} - \tilde{N}(\tilde{P}\tilde{u}^{(k)} - \tilde{\phi}), \quad k \geq 0$$

for the solution of $(\tilde{\mathrm{AS}})$ characterized by some invertible $\tilde{N}$. Applying R and denoting $u^{(k)} = R\tilde{u}^{(k)}$, we can associate a corresponding iteration method for **(AS)**:

$$u^{(k+1)} = u^{(k)} - R\tilde{N}R^*u^{(k)} + R\tilde{N}\tilde{\phi}, \quad k \geq 0.$$

Thus, playing around with the different equivalent formulations of the uniquely solvable variational problem **(VP)**, we have some more freedom when investigating iterative solvers. Indeed, the three choices

$$\tilde{N}_{JR} = \omega\tilde{I} \quad , \qquad \tilde{N}_{SOR} = (\frac{1}{\omega}\tilde{I} + \tilde{L})^{-1}$$

and

$$\tilde{N}_{SSOR} = (\frac{1}{\omega}\tilde{I} + \tilde{U})^{-1}(\frac{2}{\omega}\tilde{I} - \tilde{D})(\frac{1}{\omega}\tilde{I} + \tilde{L})^{-1}$$

lead exactly to the three Schwarz methods in V (as iterative methods in $\tilde{V}$ they are identical, up to diagonal scaling, with the classical Jacobi-Richardson (=JR), SOR, and SSOR schemes, see [Ha2]). For the additive method, this relationship is obvious since

$$I - R\tilde{N}_{JR}R^* = I - \omega RR^* = I - \omega P = \sum_{j=0}^{J}(I - \omega T_j) \equiv M_{JR}.$$

For the SOR-method we will show that indeed

$$I - R\tilde{N}_{SOR}R^* = I - R(\frac{1}{\omega}\tilde{I} + \tilde{L})^{-1}R^* = (I - \omega T_J)\dots(I - \omega T_0),$$

the latter expression coinciding with M_{SOR}. Since, analogously,

$$I - R(\frac{1}{\omega}\tilde{I} + \tilde{U})^{-1}R^* = (I - \omega T_0)\ldots(I - \omega T_J)\,,$$

the result for the third case, the symmetric SOR, follows by standard transformations. Note that we have $M_{SSOR} = M^*_{SOR}M_{SOR}$ as in the matrix case [Ha2] which will be used below in the estimations of the convergence rates for the multiplicative versions.

We verify the above identity for the multiplicative Schwarz scheme. Since $\tilde{L}$ is a $(J+1)$-dimensional, strictly lower triangular operator matrix, we have (as in the usual matrix case)

$$(\tilde{I} + \omega\tilde{L})^{-1} = \sum_{r=0}^{J}(-1)^r\omega^r\tilde{L}^r.$$

After substitution, we get

$$I - R\tilde{N}_{SOR}R^* = I - \sum_{r=0}^{J}(-1)^r\omega^{r+1}R\tilde{L}^r R^*\,.$$

For the components of the operators $\tilde{L}^r R^* : V \to \tilde{V}$ we show by induction in r that

$$(\tilde{L}^r R^*)_j = \begin{cases} 0 & , & j < r \\ \sum_{0 \le j_1 < \ldots < j_r < j} T_j T_{j_r}\ldots T_{j_1} & , & j \ge r \end{cases}$$

(for $r = 0$ this is the definition of the adjoint operator R^*, the rest is a simple substitution using $L_{ij} = P_{ij} = T_i|_{V_j}$ for $0 \le j < i \le J$ and $L_{ij} = 0$ elsewhere). Thus, after applying R we arrive at

$$R\tilde{L}^r R^* = \sum_{0 \le j_1 < \ldots < j_r < j_{r+1} \le J} T_{j_{r+1}}T_{j_r}\ldots T_{j_1}\,, \quad r = 0,\ldots,J\,.$$

This yields the above identity for M_{SOR} since the right-hand side leads to the same expressions:

$$(I - \omega T_J)\ldots(I - \omega T_0) = I - \sum_{r=0}^{J}(-1)^r\omega^{r+1}\sum_{0 \le j_1 < \ldots < j_{r+1} \le J} T_{j_{r+1}}\ldots T_{j_1}\,.$$

With these preparations at hand we can give a short proof of the main abstract convergence result for the Schwarz methods, see [Xu1, Ys1, GO2] for more details and history. We use the abbreviation $\|\cdot\|_a$ for the energy norm in V associated with the bilinear form $a(\cdot,\cdot)$ and for the corresponding operator norm.

Theorem 18. (Convergence of the Schwarz methods) Let V be finite-dimensional, equipped with the s.p.d. form $a(\cdot,\cdot)$ and with a subspace splitting $(\mathbf{D})$ into a finite number of subspaces V_j with s.p.d. forms $b_j(\cdot,\cdot)$, $j = 0,\ldots,J$.
a) The additive Schwarz method converges iff $0 < \omega < 2/\lambda_{\max}$, with the asymptotical convergence rate

$$\rho_{JR} = \|M_{JR}\|_a = \max\{|1 - \omega\lambda_{\max}|, |1 - \omega\lambda_{\min}|\} \ .$$

The optimal rate will be achieved for $\omega^* = 2/(\lambda_{\max} + \lambda_{\min})$:

$$\rho_{JR}^* = \min_{0<\omega<2/\lambda_{\min}} \rho_{JR} = 1 - \frac{2}{1+\kappa} \ .$$

A corresponding result holds for the cg-method directly applied to $(\mathbf{AS})$.
b) For the multiplicative Schwarz algorithms, convergence is guaranteed for $0 < \omega < 2/\omega_1$ where

$$\omega_1 = \lambda_{\max}(\tilde{D}) = \max_{0\neq u_j \in V_j\,,\,j=0,\ldots,J} \frac{a(u_j,u_j)}{b_j(u_j,u_j)} \ .$$

The asymptotical convergence rates can be estimated by

$$\rho_{SOR}^2 \leq \|M_{SOR}\|_a^2 = \|M_{SSOR}\|_a^2 \leq 1 - \frac{\omega(2 - \omega\omega_1)\lambda_{\min}}{\|\tilde{I} + \omega\tilde{L}\|_{\tilde{a}}^2} \ ,$$

and the optimal rates by

$$(\rho_{SOR}^*)^2 \leq \rho_{SSOR}^* \leq 1 - \frac{\lambda_{\min}}{2\|\tilde{L}\|_{\tilde{a}} + \omega_1} \ .$$

Proof. We concentrate on the proof of the bound for $\|M_{SSOR}\|_a$, the other details are left up to the reader, see also [GO2]. By the above definitions and representations,

$$a(M_{SSOR}u, u) = a(u,u) - a(R(\tfrac{1}{\omega}\tilde{I} + \tilde{U})^{-1}(\tfrac{2}{\omega}\tilde{I} - \tilde{D})(\tfrac{1}{\omega}\tilde{I} + \tilde{L})^{-1}R^*u, u)$$

$$= a(u,u) - \tilde{a}((\tfrac{2}{\omega}\tilde{I} - \tilde{D}^*)\tilde{w}, \tilde{w})$$

where $\tilde{w} = (\tfrac{1}{\omega}\tilde{I} + \tilde{L})^{-1}R^*u$. If the convergence condition for the multiplicative methods is fulfilled, then $(\tfrac{2}{\omega}\tilde{I} - \tilde{D})$ is s.p.d. on $\tilde{V}$ and we can continue

$$\tilde{a}((\tfrac{2}{\omega}\tilde{I} - \tilde{D}^*)\tilde{w}, \tilde{w}) \geq (\tfrac{2}{\omega} - \omega_1)\tilde{a}(\tilde{w}, \tilde{w})$$

$$\geq (\tfrac{2}{\omega} - \omega_1)\|(\tfrac{1}{\omega}\tilde{I} + \tilde{L})\|_{\tilde{a}}^{-2}\tilde{a}(R^*u, R^*u)$$

$$
= \ (\frac{2}{\omega} - \omega_1)\|(\frac{1}{\omega}\tilde{I} + \tilde{L})\|_{\tilde{a}}^{-2} a(Pu, u)
$$

$$
\geq \ (\frac{2}{\omega} - \omega_1)\|(\frac{1}{\omega}\tilde{I} + \tilde{L})\|_{\tilde{a}}^{-2} \lambda_{\min} a(u, u).
$$

Altogether, we have now the desired result for the convergence rate of the two multiplicative Schwarz schemes (recall the definition of a norm). After a simplification by using

$$
\|(\frac{1}{\omega}\tilde{I} + \tilde{L})\|_{\tilde{a}} \leq \frac{1}{\omega} + \|\tilde{L}\|_{\tilde{a}}
$$

and minimization with respect to $t = \frac{1}{\omega}$ one obtains also the bound for the optimal rates.

We add a few comments on the different constants entering the estimates of Theorem 18. The constant ω_1 does not cause serious problems. In the case when $b_j(\cdot, \cdot)$ is induced by the form $a(\cdot, \cdot)$ (exact solvers for the subproblems) we have $\omega_1 = 1$, in the general case, still $\omega_1 \leq \lambda_{\max}$ holds true. Eventually estimates can be obtained by comparing the forms separately on each subspace.

Estimates of $\lambda_{\min}$ depend on knowledge about good decompositions of elements from V into building blocks from the subspaces V_j. Especially for this task, the methods of approximation theory should be helpful. Examples are given below.

What concerns estimates for $\lambda_{\max}$ and $\|\tilde{L}\|_{\tilde{a}}$, one of the basic techniques is the use of so-called strengthened Cauchy-Schwarz inequalities. Following [Xu1], they are stated in the following abstract form: there exist positive constants γ_{ij} so that

$$
a(u_i, v_j) \leq \gamma_{ij}\sqrt{b_i(u_i, u_i)}\sqrt{b_j(v_j, v_j)} \qquad \forall\, u_i \in V_i \ \forall v_j \in V_j,
$$

for all $i, j = 0, \ldots, J$. Without loss of generality, we may assume that $\gamma_{ij} = \gamma_{ji}$. Let K_1 denote the largest eigenvalue ($=$ spectral radius) of the symmetric matrix $\{\gamma_{ij}\}$ formed by these constants. Note that a trivial estimate of K_1 can often be given by the number of subspaces involved, this estimate can be improved by introducing the notion of the colouring number of a splitting, see [Xu1]. It is easy to see that K_1 is an upper bound for $\lambda_{\max}$, $\|\tilde{L}\|_{\tilde{a}}$, and ω_1, simultaneously. We show this for $\|\tilde{L}\|_{\tilde{a}}$, the other cases are simpler and left to the reader. Indeed, for arbitrary $\tilde{u}, \tilde{v} \in \tilde{V}$ we have

$$
\tilde{a}(\tilde{L}\tilde{u}, \tilde{v}) = \sum_{i=1}^{J}\sum_{j=0}^{i-1} b_i(T_i u_j, v_i)
$$

$$
\cdot = \sum_{i=1}^{J}\sum_{j=0}^{i-1} a(u_j, v_i)
$$

$$\leq \sum_{i=0}^{J}\sum_{j=0}^{J} \gamma_{ij} \sqrt{b_j(u_j, u_j)} \sqrt{b_i(v_i, v_i)}$$

$$\leq K_1 \sqrt{\tilde{a}(\tilde{u}, \tilde{u})} \sqrt{\tilde{a}(\tilde{v}, \tilde{v})} \, ,$$

it remains to put $\tilde{v} = \tilde{L}\tilde{v}$ and to use the norm definition.

In case that sharp Cauchy-Schwarz inequalities are not easy to obtain, the following general bound for the lower triangular part of a s.p.d. operator matrix can be helpful (see [GO2]):

$$2\|\tilde{L}\|_{\tilde{a}} \leq \log_2(2J)\|\tilde{P}\|_{\tilde{a}} = \log_2(2J)\lambda_{\max} \, .$$

Substituting this inequality into the estimates of Theorem 18 b), we get one more quantitative bound for the multiplicative Schwarz methods which is slightly worse than the optimal convergence rates of the additive algorithm:

$$(\rho_{SOR}^*)^2 \leq \rho_{SSOR}^* \leq 1 - \frac{1}{\log_2(4J)\kappa} \, .$$

As has recently been shown in [Os11], the logarithmical factor in this estimate is necessary, the examples in [Os11] are given for the case of the classical SOR and SSOR methods for s.p.d. linear systems in $\mathbf{R}^n$ which are indeed particular cases of the above multiplicative Schwarz schemes. Even if in practice the multiplicative versions are usually better with respect to convergence rates than the additive method, the latter estimate states an important theoretical fact: for *any* finite subspace splitting $(\mathbf{D})$, the properly scaled multiplicative methods converge at a rate depending on the condition number of the additive Schwarz operator, and at most logarithmically on the number of subspaces. Good preconditioning via the additive Schwarz formulation plus correct scaling automatically implies fast convergence of the multiplicative Schwarz algorithm if the number of subspaces is moderate! For some inverse result, compare [BW1].

The effect of individual scaling (this is replacing ω by ω_j, $j = 0, 1, \ldots, J$) in the above algorithm as well as for improving the bounds for the additive Schwarz operator (this is replacing $b_j(\cdot, \cdot)$ by $b_j(\cdot, \cdot)/\omega_j$) has not yet been studied in a satisfactory way. Note that constant scaling ($\omega_j = \omega$) does not change the condition of the additive operator though it substantially influences the behavior of the multiplicative algorithm.

As a whole, we have now a very compact basic theory, with clear relationships between the additive and multiplicative Schwarz methods. There are certainly shortcomings: the above estimates in the multiplicative case are still very rough and pessimistic, and are not explaining some traditional topics of multigrid theory like the dependence of the rates on the number of pre- and post-smoothing steps (compare [Ys1, Sd] in this respect). Note that there have been proposed generalizations of the above abstract theory to mildly indefinite and/or non-symmetric problems, cf. [Xu4, CX, CW, BL2, Wa1, Wa2].

Throughout the remainder of our notes, we will exlusively concentrate on condition number estimates for the additive version (with the good feeling of dealing implicitely with the multiplicative schemes as well, cf. Theorem 18 and the remarks following it). According to Theorem 16, this reduces to the proof, with best possible constants, of the norm equivalence **(NE)**, now written for finitely many subspaces as

$$a(u,u) \approx \inf_{u_j \in V_j \,:\, u = \sum_{j=0}^{J} u_j} \sum_{j=0}^{J} b_j(u_j, u_j) \,.$$

One method to get such estimates is to look at the Besov space characterizations from approximation theory (Theorem 6 and, in a more applicable form, Theorem 15). They provide us with stable multilevel splittings into infinitely many subspaces for infinite-dimensional Sobolev spaces from which finite, uniformly stable splittings can be obtained for computational purposes. Decompositions of the full Sobolev space are useful also for error estimation, see section 5. Alternative (sufficient) conditions which are closer in spirit to the tradition in multigrid theory to rely on approximation and smoothing properties are given by many authors, see [BW1, BW2, Xu1, Ys1, Sd] and the references cited therein.

Last but not least, one can produce many more good splittings from a given one by *refinement* and *clustering*. By this we mean the following procedures. Given a splitting **(D)** with appropriate auxiliary forms $b_j(\cdot, \cdot)$ and an estimate for the corresponding condition number $\kappa \equiv \mathrm{cond}(P)$ of the additive Schwarz operator, we assume that all or some of the V_j are further splitted into sums of their closed subspaces $V_{j,i}$ equipped with s.p.d. forms $b_{j,i}(\cdot, \cdot)$:

$$\{V_j, b_j\} = \sum_i \{V_{j,i}, b_{j,i}\} \,.$$

Figure 13 shows this schematically. Suppose that we can control the numbers

$$\Lambda = \max_j \lambda_{\max}(P_j) \,, \quad \lambda = \min_j \lambda_{\min}(P_j) \,,$$

where the additive Schwarz operators P_j are defined with respect to the above splittings for the variational problem on V_j with the bilinear form $b_j(\cdot, \cdot)$. Then, for the condition number of the additive Schwarz operator P_{ref} of the refined

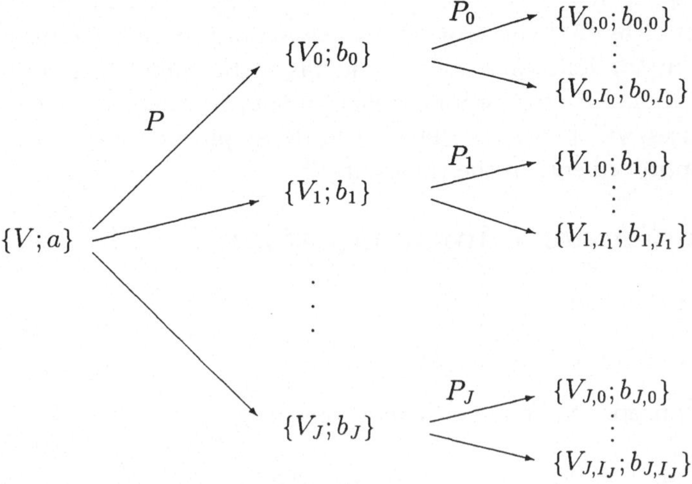

Figure 13. Refinement and clustering

splitting

$$\{V;a\} = \sum_{j=0}^{J} \sum_{i} \{V_{j,i}, b_{j,i}\}$$

we have the bound

$$\mathrm{cond}(P_{ref}) \leq \frac{\Lambda}{\lambda}\,\mathrm{cond}(P)\ .$$

Conversely, if there is some knowledge about $\mathrm{cond}(P_{ref})$ then one can use

$$\mathrm{cond}(P) \leq \frac{\Lambda}{\lambda}\,\mathrm{cond}(P_{ref})\ .$$

These two obvious relationships (look at the corresponding inequalities in **(NE)**) are the basis for refinement and clustering techniques, respectively, which will be used below without further being mentioned. Subspace *selection* will be discussed in connection with adaptive refinement in 4.2.2.

Anyhow, to find optimal bounds for a given splitting (or to find a splitting with good bounds) still remains a delicate task requiring insight into the given variational problem, and a good understanding of the algorithmical aspect, too. Not every splitting is efficient, minimizing the condition number of P sometimes leads to an inadequate increase of computational complexity in the corresponding algorithms. We do not make any attempt to define an efficiency measure but it is clear that the abstract discussion of preconditioning via subspace splittings in the following subsections has to be complemented by such kind of considerations. At the end, practical implementations on modern computer architectures will judge the theoretical developments, and lead to new questions and improvements of the theory itself.

4.2   Second-order elliptic equations

We consider the problem

$$-\nabla(p\nabla u) + qu = f \qquad \text{in} \quad \Omega$$

equipped with appropriate boundary conditions, e.g.,

$$u = g \quad \text{on } \Gamma \subset \partial\Omega \, ; \qquad \frac{\partial u}{\partial n} = h \quad \text{on } \partial\Omega\backslash\Gamma \, .$$

As before, Ω is a bounded polyhedral domain in $\mathbf{R}^d$.

Our aim is to apply the results of 3.4 in conjunction with Theorem 16. We will assume that our problem is $H^1_{D=\Gamma}$-elliptic, with constants in the equivalence relation

$$a(u, u) = (p\nabla u, \nabla u)_{L_2} + (qu, u)_{L_2} \approx \|u\|^2_{H^1} \qquad \forall\, u \in H^1_{D=\Gamma}(\Omega)$$

that depend on the coefficient bounds for p, q, and on the domain. Thus, stable splittings for $H^1_{D=\Gamma}$ (with its standard scalar product as bilinear form) will yield splittings for the variational problem corresponding to the above elliptic boundary value problem. The point is that a comparison of Theorem 15 and Theorem 16 immediately indicates one possible candidate for such splittings in the case of multilevel-structured f.e. spaces $V_J \subset H^1_{D=\Gamma}$. Modifications which include the theory of the celebrated Bramble-Pasciak-Xu (BPX) preconditioner [BX1], Zhang's MDS [Zh1, Wa2], and Yserentant's hierarchical basis method [Ys2] as well as others are discussed in 4.2.1. In 4.2.2 we deal with the case of triangulations resulting from nested refinement. Finally, 4.2.3 contains some brief discussion on the dependence of the coefficients of the differential operator, the case of parabolic problems and other aspects.

4.2.1   The basic multilevel preconditioners

The first result applies to C^0 Lagrange elements, i.e. we can fix $r = r' = 0$, $m = m' \geq 2$ (actually any of the schemes that meet the requirements of Theorem 15 will work but there is no plausibility in taking complicated C^1 elements for discretizing a second order elliptic equation). We use the notation and the assumptions concerning $\{T_j\}$ $(a = 2)$ and Γ introduced in sections 2 and 3.

Let $\mathcal{N}_j = \{P_{j,i}\}$ be the set of nodal points in $\bar\Omega\backslash\Gamma$, in the case considered to each $P_{j,i} \in \mathcal{N}_j$ there corresponds a unique nodal basis function $N_{j,i} \in V_j \equiv V(T_j)|_{D=\Gamma}$, the one-dimensional subspace spanned by this function is denoted by $V_{j,i}$. We suppose that $\mathcal{N}_0 \neq \emptyset$, and set $\mathcal{N}_{-1} = \emptyset$.

We may introduce another basis in each V_j which is called hierarchical basis (for a somewhat more general definition what might be called hierarchical basis in f.e. multilevel schemes, see [Df1]):

$$\{H_{j,i}\} \equiv \{N_{j,i}\,,\ P_{j,i} \in \mathcal{N}_j\backslash\mathcal{N}_{j-1}\,,\ j = 0,\ldots,J\} \quad .$$

Moreover, one might think about a prewavelet-like basis

$$\{w_{j,i}\} \equiv \{w_{j,i} \in W_j \equiv V_j \ominus V_{j-1}\,,\ P_{j,i} \in \mathcal{N}_j\backslash\mathcal{N}_{j-1}\,,\ j = 0,\ldots,J\} \quad .$$

For general f.e. multilevel schemes $\{V_j\}$ over domains, the construction of such a partially orthogonalized basis of localized functions $w_{j,i}$ forming a Riesz basis in each orthogonal complement W_j (with L_2 condition uniformly bounded for $j \to \infty$) is an interesting open theoretical problem. But in some particular cases, e.g., for the d-dimensional cube, prewavelet systems are available by simple tensor-product constructions from the univariate wavelet theory, see the examples in 2.1.3. More economic multivariate prewavelet constructions, i.e. yielding $w_{j,i}$ with smaller support, are certainly possible. In Figure 14, an example is indicated for the case of linear f.e. constructions over a uniform type-I-triangulation of the unit square found in [KO] (modifications near the boundary are still necessary as in the univariate case). The numbers given in

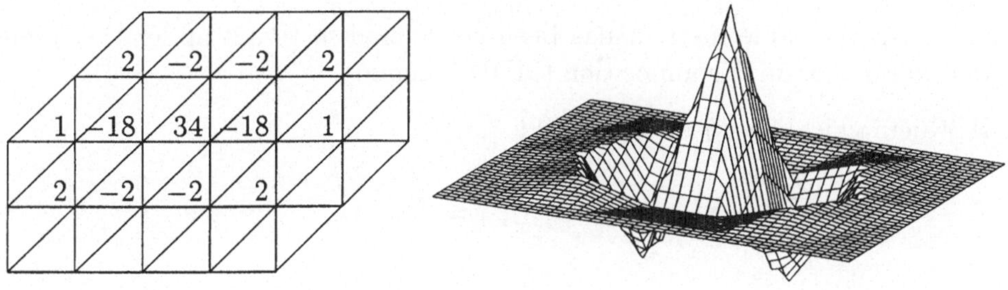

Figure 14. A linear prewavelet for type-I-triangulations

Figure 14 are the 13 non-zero coefficients scaled to integer values in the nodal basis decomposition of a generic interior $w_{j,i}$. This example exhibits minimal support, i.e. any nontrivial $w \in W_j$ on this type of triangulation has a support of ≥ 10 triangles from $\mathcal{T}_{j-1}$ (there are, however, possibilities to slightly reduce the number of non-zero coefficients in the corresponding mask, see [KO]). Note that the nodal basis functions $N_{j-1,i}$ used in the other (non-wavelet) splittings can be represented by 7 nodal basis functions from V_j. Thus, the partial orthogonality has to be paid by an enlargement by a factor of ≈ 2 of the size of the masks characterizing the intergid transfer operations in the multilevel scheme to be described. See section 4.4 for an example where a prewavelet decomposition might be useful.

Theorem 19. Given a symmetric elliptic problem in $H^1_{D=\Gamma}$, with $a(\cdot,\cdot)$ denoting the corresponding bilinear form, let $\{V_j = V(\mathcal{T}_j)|_{D=\Gamma}\}$ be a C^0 Lagrange element multilevel scheme with underlying regular dyadic refinement ($a = 2$). Assume $V_0 \neq \emptyset$. Then the following splittings have the asymptotically sharp condition number estimates as indicated:

- $V = \sum_{j=0}^{J} V_j$, $b_j(\cdot,\cdot) = a(\cdot,\cdot)$: $\kappa = J + 1$

- $V = \sum_{j=0}^{J} V_j$, $b_j(\cdot,\cdot) = 2^{2j}(\cdot,\cdot)_{L_2}$: $\kappa = O(1)$, $J \to \infty$

- BPX-method [BX1]

$$V = \sum_{j=0}^{J} \sum_{P_{j,i} \in \mathcal{N}_j} V_{j,i} , \; b_{j,i}(\cdot,\cdot) = 2^{2j}(\cdot,\cdot)_{L_2} \quad : \quad \kappa = O(1) , \; J \to \infty$$

- MDS(multilevel diagonal scaling)-method [Zh1]

$$V = \sum_{j=0}^{J} \sum_{P_{j,i} \in \mathcal{N}_j} V_{j,i} , \; b_{j,i}(\cdot,\cdot) = a(\cdot,\cdot) \quad : \quad \kappa = O(1) , \; J \to \infty$$

(actually, the same method has been considered in [Wa2] under the name multilevel domain decomposition (MDD-scheme)).

- HB(hierarchical basis)-method [Ys2]

$$V = \sum_{j=0}^{J} \sum_{P_{j,i} \in \mathcal{N}_j \backslash \mathcal{N}_{j-1}} V_{j,i} , \; b_{j,i}(\cdot,\cdot) = 2^{2j}(\cdot,\cdot)_{L_2} \quad :$$

$$\kappa = \begin{cases} O(1) , & d = 1 \\ O((J+1)^2) , & d = 2 \\ O(2^{(d-2)J}) , & d \geq 3 \end{cases} , \; J \to \infty$$

(other choices of the bilinear forms are possible [Ys3]).

- Prewavelet splitting: Suppose that a prewavelet-like system $\{w_{j,i}\}$ exists with the properties as described above, and denote $W_{j,i} = \operatorname{span} w_{j,i}$.

$$V = \sum_{j=0}^{J} \sum_{P_{j,i} \in \mathcal{N}_j \setminus \mathcal{N}_{j-1}} W_{j,i} \, , \quad b_{j,i}(\cdot,\cdot) = 2^{2j}(\cdot,\cdot)_{L_2} \quad :$$

$$\kappa = O(1) \, , \quad J \to \infty \, .$$

The constants in the bounds depend on Ω, Γ, $\mathcal{T}_0$, the degree of the Lagrange elements, and the bounds in

$$a(u,u) \approx \|u\|_{H^1}^2 \quad , \qquad u \in H^1_{D=\Gamma}(\Omega) \, .$$

Proof. Besides the first splitting and the HB splitting, the results are contained in Theorem 15, $s = 1$, $r = 0$, if one consults Theorem 16. The theory of the MDS-splitting is a corollary to that of the BPX-splitting since $a(N_{j,i}, N_{j,i}) \approx \|N_{j,i}\|_{H^1}^2 \approx 2^{2j}\|N_{j,i}\|_{L_2}^2$ (for an independent proof, see [Zh1]). The prewavelet case is also straightforward: if

$$u = \sum_{j=0}^{J} \sum_{i} d_{j,i} w_{j,i} \in V_J$$

is the prewavelet decomposition of u then

$$Q_j u - Q_{j-1} u = \sum_{i} d_{j,i} w_{j,i} \in W_j \, , \quad j = 0, \ldots, J \, ,$$

and it remains to apply the uniform Riesz basis property of the prewavelets in W_j in the corresponding norm equivalence of Theorem 15 (actually, we apply all the time the refinement trick mentioned at the end of the previous section).

For the HB-splitting, consult Theorem 12, with $p = 2$, $s = 1$. Examples showing the sharpness of the asymptotics are given in [Ys1] $(d = 2)$ and [Os5, On] $(d \geq 3)$.

The first splitting is elementary: For any $u = \sum_{j=0}^{J} u_j \in V_J$ with $u_j \in V_j$, $j = 0, \ldots, J$, we have

$$a(u,u) = \left(\sum_{j=0}^{J} \sqrt{a(u_j,u_j)} \right)^2 \leq (J+1) \sum_{j=0}^{J} a(u_j,u_j)$$

with equality, if $u_j = u_0 \in V_0 \subset V_j$, $j = 0, \ldots, J$. This shows that $a(u,u) \leq (J+1) \cdot \|\|u\|\|^2$ is sharp. On the other hand, $a(u,u) \geq \|\|u\|\|^2$ (look at the decomposition given by $u_j = 0$ for $j < J$ and $u_J = u$). The sharpness of this trivial lower bound can be shown for u from $\tilde{W}_J$, the a-orthogonal complement of V_{J-1} in V_J.

Note that the first splitting was included only for the sake of completeness - it shows that without changing the bilinear forms (compare the second splitting), the monotonicity of the multilevel subspace structure introduces too much overlap resulting in a detoriation of the condition number if $J \to \infty$. Decompositions into direct sums of subspaces also do not lead, as a rule, to asymptotically optimal condition numbers (HB-method if $d \geq 2$), or are more involved (prewavelet methods). The prewavelet method may be viewed as an attempt to explicitely implement the L_2 orthogonality which is implicitly used in the theory of the BPX-method. In the H^1-setting considered here, it seems to us that the wavelet approach will not bring any advantage in computational efficiency compared to BPX-like methods (but compare the discussion in [Ja, JL]). Looking at the computational formula corresponding to one preconditioning step in the BPX-method

$$P_{BPX}u = \sum_{j=0}^{J} \sum_{P_{j,i} \in \mathcal{N}_j} \frac{a(u, N_{j,i})}{2^{2j}(N_{j,i}, N_{j,i})} N_{j,i}, \quad u \in V_J,$$

resp. the hypothetical prewavelet method

$$P_{PW}u = \sum_{j=0}^{J} \sum_{P_{j,i} \in \mathcal{N}_j \setminus \mathcal{N}_{j-1}} \frac{a(u, w_{j,i})}{2^{2j}(w_{j,i}, w_{j,i})} w_{j,i}, \quad u \in V_J,$$

one sees that the latter requires more work since the masks corresponding to the level-to-level computations in the prewavelet transform case are substantially larger then the masks for the projections to finer levels in the nodal basis case (for a two-dimensional application the factor is approximately 2). A closer look to the pyramid scheme [Db] governing all wavelet and prewavelet calculations shows that things are even worse if one still uses the nodal basis discretization matrix: the prewavelet preconditioning V-cycle automatically contains a dummy BPX-cycle plus the work necessary for applying the prewavelet masks on all levels! Besides a lot of additional problems not yet solved for the prewavelet method, there is at the moment no numerical evidence concerning the expected improvement of preconditioning. We address this problem of actual theoretical and practical interest to the reader. Note, however, that there is much more activity about using the wavelet concept for p.d.e. solving than what we have briefly described here (especially, operator adapted wavelet schemes might have some chances even for practical codes, see [Dh, DP1, DP2, Dk1, Dk2, Ku, Bl, CDJ, JaS]).

We end this section with applications of the clustering trick. First of all, if we collect all subspaces with level number $\leq j_0$ into the one subspace V_{j_0} ($0 \leq j_0 < J$) and use the BPX-part of Theorem 19 once again (this time with J replaced by j_0) we see that the splitting

$$V = V_{j_0} + \sum_{j=j_0+1} \sum_{P_{j,i} \in \mathcal{N}_j} V_{j,i} \, , \qquad b_{j_0}(\cdot,\cdot) = a(\cdot,\cdot) \, , \; b_{j,i}(\cdot,\cdot) = 2^{2j}(\cdot,\cdot)_{L_2}$$

possesses also an $O(1)$ bound for the condition of the additive Schwarz operator which is independent of j_0 and J. As before, the scaled L_2 scalar products can be replaced by the original form $a(\cdot,\cdot)$ (MDS variant). In practice, the corresponding preconditioners are often advantegeous since they include the exact solution of a coarse grid problem corresponding to the level j_0. It also allows us to treat domains of more complex shape under the assumption that the partition $\mathcal{T}_{j_0}$ resolves the geometry of the domain. The codes PLTMG [Ba] and Kaskade [DLY, Ln, Ro] use this technique. Note that, for the hierarchical basis preconditioner, one is recommended to repeat the proof of Theorem 12 which gives somewhat better bounds (e.g., $O((J - j_0)^2)$ in the two-dimensional case) than the above argument.

We give a second example. For linear elements, Griebel [Gr4] introduced the so-called point-block preconditioner which improves the parallelism of the MDS-BPX preconditioner. The underlying splitting is

$$V = \sum_{j=0}^{J} \sum_{P_{j,i} \in \mathcal{N}_j \backslash \mathcal{N}_{j-1}} \hat{V}_{j,i} \, , \quad b_{j,i}(\cdot,\cdot) = a(\cdot,\cdot) \, ,$$

where $\hat{V}_{j,i}$ denotes the $(J - j + 1)$-dimensional subspace spanned by all basis functions $N_{l,i'}$ with $l \geq j$ corresponding to the same grid point $P_{j,i} \in \mathcal{N}_j \backslash \mathcal{N}_{j-1}$ of level j (to simplify the notation, let us order the nodes within a level j starting with the nodal points from level $j - 1$ such that $\hat{V}_{j,i} = \{\sum_{l=j}^{J} c_{l,i} N_{l,i}\}$ holds true). Even if we now have to solve small subproblems of dimension $\leq J$, the number of arithmetical operations per preconditioning step remains $O(\dim V_J)$ as for the MDS preconditioner. To show an $O(1)$ bound for the condition number it is enough to prove the norm equivalence

$$a\Big(\sum_{l=j}^{J} c_l N_{l,i}, \sum_{l=j}^{J} c_l N_{l,i}\Big) \approx \sum_{l}^{J} c_l^2 a(N_{l,i}, N_{l,i}) \, ,$$

and to use the clustering argument starting from the MDS-splitting result of Theorem 19. Let us consider triangular linear elements in R^2 (there is no change

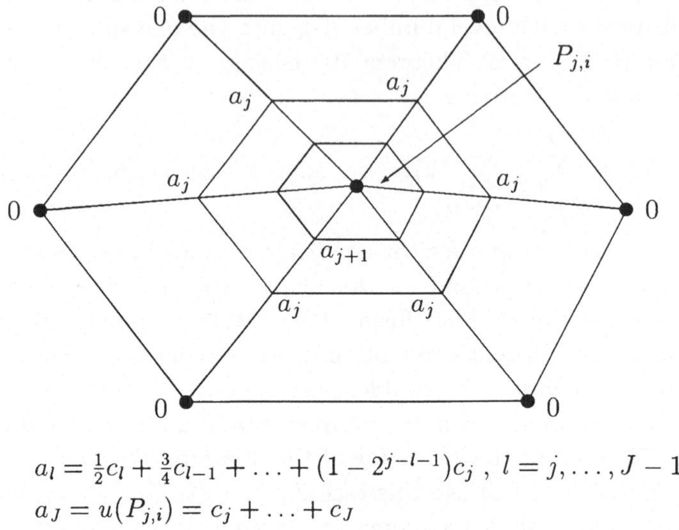

$$a_l = \tfrac{1}{2}c_l + \tfrac{3}{4}c_{l-1} + \ldots + (1 - 2^{j-l-1})c_j \ , \ l = j, \ldots, J-1$$
$$a_J = u(P_{j,i}) = c_j + \ldots + c_J$$

Figure 15. Nodal values of $u \in \hat{V}_{j,i}$

in the argument if one goes to higher dimensions, rectangles, or uses hierarchical shape functions for higher degree Lagrange elements, another more general proof has been given in the extended version of [Os15]). Figure 15 shows the support and the nodal values of an arbitrary function $u = \sum_{l=j}^{J} c_l N_{l,i} \in \hat{V}_{j,i}$. Since the squared H^1-seminorm of a linear function over a triangle is proportional to the sum of the squares of the differences of the values at the three vertices, we have from this picture

$$a(u,u) \approx \sum_{l=j}^{J}(a_l - a_{l-1})^2 \approx \sum_{l=j}^{J}\underbrace{(c_l + c_{l-1}/2 + \ldots + 2^{-(l-j)}c_j)^2}_{b_l}$$

where the values of a_l are given in Figure 15 ($a_{j-1} = 0$). Since $b_l = c_l + b_{l-1}/2$ resp. $c_l = b_l - b_{l-1}/2$ ($b_{j-1} = 0$) one immediately obtains $\sum_{l=j}^{J} b_l^2 \approx \sum_{l=j}^{J} c_l^2$ which yields the desired norm equivalence with constants independent of $u \in \hat{V}_{j,i}$ and the chosen nodal point $P_{j,i} \in \mathcal{N}_j \backslash \mathcal{N}_{j-1}$.

There are many other groups of subspaces one can collect together without destroying the condition number too much. This is important if one looks for an implementation on a particular computer architecture. We want to emphasize that, in some sense, the *best* possible splitting associated with our multilevel f.e.

scheme $\{V_j\}$ is given by

$$V_J = \sum_{j=0}^{J} \tilde{W}_j\,, \qquad \tilde{W}_j = \mathrm{Im}(P_j - P_{j-1})\,,$$

where $P_j\,:\,H^1_{D=\Gamma} \to V_j$ denote the projections with respect to the bilinear form $a(\cdot,\cdot)$:

$$a(P_j u, v_j) = a(u, v_j) \qquad \forall\, v_j \in V_j\,,\ j = 0,1,\dots\ (P_{-1} \equiv 0).$$

This is an a-orthogonal direct sum of subspaces which formally leads to an additive Schwarz operator with condition number $\kappa = 1$ if the auxiliary forms on the $\tilde{W}_j$ are simply the restrictions of $a(\cdot,\cdot)$ onto this subspace. Clearly, a straightforward use of this splitting is impossible, the solution of the subproblems is as complex as the original problem (**VP**). However, it seems to be interesting and possible to analyze multigrid preconditoners (as well as standard multigrid methods, the multiplicative version) by looking for conditions on auxiliary forms $b_j(\cdot,\cdot)$ defined on the whole V_j that give rise to good substitutes for the exact projection $P_j - P_{j-1}$ onto $\tilde{W}_j$. Possibly, this is also a way to figure out the role of pre- and post-smoothing steps in the standard multigrid method via the above abstract setting. We refer to the recent papers [BP2, BP3, BW1, BW2], see also [Xu1, Ys1, Sd]. In addition, the construction of additive splittings of $\{\tilde{W}_j; a(\cdot,\cdot)\}$ into local building blocks may lead to another prewavelet-like method which looks quite attractive and should be compared with the existing algebraic multigrid methods.

Till now, the considerations of this subsections have been restricted to C^0 Lagrange elements. We finish with some remarks on how to construct preconditioners for serendipity elements, or for C^0 Hermite elements. The same idea, i.e. the use of coarse grid spaces of linear finite elements, works for higher degree Lagrange elements, too. Consider some subspace of triangular elements $V \equiv V(\mathcal{T}_J)$ such that it contains the space V_J of linear f.e. over the same partition. Then the splitting

$$V = \sum_{j=0}^{J-1} V_j + V$$

where the subspaces are equipped with the scaled L_2 scalar products as usual (accordingly, the auxiliary form on V is $2^{2J}(\cdot,\cdot)_{L_2}$), leads to a preconditioner with an $O(1)$ bound of the condition numbers (below we need additionally the inverse assumption for V which is satisfied for the finite element types mentioned above). Indeed, let $\mathcal{Q}_j$, denote the quasi-interpolants projecting onto V_j, $j < J$.

Looking at the proof of the Jackson-type estimate of Theorem 11, and the specification given in Theorem 15, we see that

$$2^{2J}\|u - \mathcal{Q}_{J-1}u\|_{L_2}^2 \leq C\|u\|_{A_{2,2}^1(\{V_j, D=\Gamma\})}^2 \leq Ca(u, u)$$

and

$$a(\mathcal{Q}_{J-1}u, \mathcal{Q}_{J-1}u) \approx \sum_{j=0}^{J-1} 2^{2j}\|\mathcal{Q}_j u - \mathcal{Q}_{j-1}u\|_{L_2}^2$$
$$\leq \sum_{j=0}^{\infty} 2^{2j}\|\mathcal{Q}_j u - \mathcal{Q}_{j-1}u\|_{L_2}^2 \approx a(u, u).$$

According to Theorem 16, this gives the estimate for the spectrum of the additive Schwarz operator from below. For the upper bound, let $u = \sum_{j=0}^{J-1} u_j + v$ ($u_j \in V_j$, $v \in V$) be an arbitrary decomposition of $u \in V$. Then

$$a(u, u) \leq 2(a(\sum_{j=0}^{J-1} u_j, \sum_{j=0}^{J-1} u_j) + a(v, v)) \leq C\left(\sum_{j=0}^{J-1} 2^{2j}\|u_j\|_{L_2}^2 + 2^{2J}\|v\|_{L_2}^2\right)$$

where Theorem 15 (linear f.e. case) and the inverse inequality for the H^1 norm of $v \in V$ have been used (the latter can be checked for the V under consideration directly, without using Theorem 11). This result shows once again that, besides some epsilons, the case of linear finite elements contains all difficulties and beauties which are important for C^0 finite element constructions. The epsilons are, e.g., higher approximation order if the functions to be approximated belong to higher order smoothness classes.

4.2.2 Nested refinement

This point has been of interest since the beginning of multigrid developments. After the efforts of many mathematicians, especially [Ys2, BX1, BP3, DK, BY], it became clear that most of the preconditioners considered in 4.2.1 can be run successfully for non-quasiuniform partitions resulting from an adaptive (or a priori determined) multilevel refinement process. The basic restriction on the refinement procedure (which is dictated also by the data structures used in programs) is that it remains nested, i.e. subdomains (triangles, tetrahedra, rectangles,...) of level j which are not candidates for further refinement will never be touched on in the future. We do not describe the variants of this basic idea which is used in many of the currently available programs for adaptive solution of elliptic equations, see [Ba, DLY, Bä, BE1].

What we want to show, however, is which kind of argument reduces the theory of multilevel preconditioners for the nested refinement case to the case of regular dyadic refinement considered till now. Our assumption concerning the sequence of adaptively refined partitions $\{T_j\}$ is as follows: There exists another increasing sequence of uniformly quasi-uniform and regular partitions $\{T_j^*\}$ such that $T_0^* = T_0$, $h(T_j^*) \approx 2^{-j}$, T_j is contained in T_j^*, and coincides with the latter for all those subdomains which were subdivided when going from level $j-1$ to j in the refinement process. The union of these subdomains will be denoted by Ω_j ($\Omega_0 = \Omega$). We also require that on $\Omega \backslash \Omega_j$ all partitions T_l with $l \geq j-1$ are identical, i.e. refinement of level $\geq j$ does not occur outside Ω_j, and $\Omega_j \subset \Omega_{j-1}$, $j \geq 1$. This latter requirement is actually the version of nested refinement we will make use of in the proof. Figure 16 schematically shows the compatibility of the above assumption with the typical red-green refinement used in PLTMG or Kaskade. Denote by $V_j = V(T_j)$ resp. $V_j^* = V(T_j^*)$ subspaces of linear finite

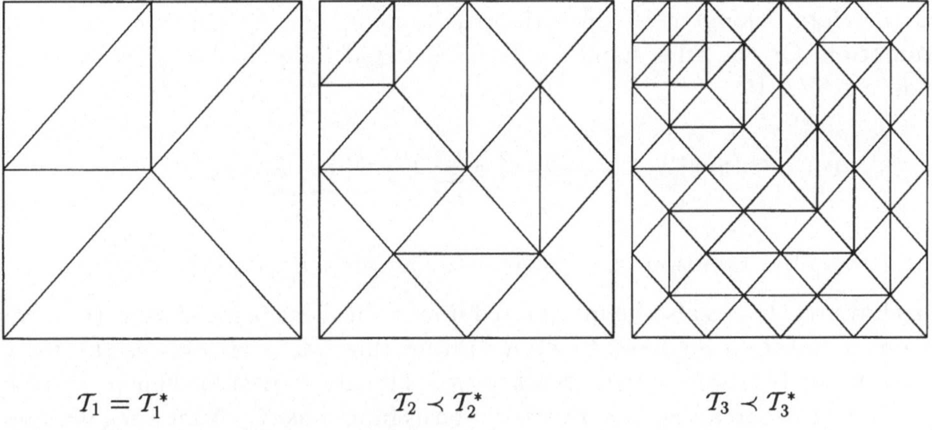

$T_1 = T_1^*$ $T_2 \prec T_2^*$ $T_3 \prec T_3^*$

Figure 16. Nested refinement

element functions (other C^0 elements will do as well, the incorporation of the Dirichlet conditions on Γ is assumed without explicitly notifying this), and set $\tilde{V}_j = \text{span}\,\{N_{j,i}\,|\,\text{span}\,N_{j,i} \subset \Omega_j\}$. To see that the splittings

$$\{V_J; a\} = \sum_{j=0}^{J} \{\tilde{V}_j; 2^{2j}(\cdot,\cdot)_{L_2}\} = \sum_{j=0}^{J} \sum_{i\,:\,\text{supp}\,N_{j,i} \subset \Omega_j} \{V_{j,i}; 2^{2j}(\cdot,\cdot)_{L_2}\}$$

(in the last splitting we can also replace the forms on the $V_{j,i}$ by $a(\cdot,\cdot)$) give rise to uniformly bounded condition numbers for the corresponding additive Schwarz operators, we can use Theorem 19 for the sequence V_j^*, together with a special quasi-interpolant construction. Indeed, since $\tilde{V}_j \subset V_j^*$ we get for any

$u \in V_J$

$$a(u,u) \approx \inf_{u_j^* \in V_j^* \,:\, u=\sum u_j^*} \sum_{j=0}^{J} 2^{2j} \|u_j^*\|_2^2 \le \inf_{\tilde{u}_j \in \tilde{V}_j \,:\, u=\sum \tilde{u}_j} \sum_{j=0}^{J} 2^{2j} \|\tilde{u}_j\|_2^2 \,.$$

The lower estimate for the spectrum follows if we decompose $u \in V_J \subset V_J^*$ according to

$$u = \mathcal{Q}_0^* u + \sum_{j=1}^{J} (\mathcal{Q}_j^* u - \mathcal{Q}_{j-1}^* u)$$

where the supporting triangles of the coefficient functionals $\lambda_{j,i}^*(\cdot)$ in the definition of the quasi-interpolants $\mathcal{Q}_j^* : L_1(\Omega) \to V_j^*$ (see subsection 2.1.1) are chosen outside Ω_j whenever $\operatorname{supp} N_{j,i}^*$ is not completely contained in Ω_j. This ensures by construction that the differences $\mathcal{Q}_j^* u - \mathcal{Q}_{j-1}^* u$ vanish outside Ω_j for all $j = 1, \ldots, J$ if u belongs to V_J. Thus, these differences have support in Ω_j, and therefore belong to $\tilde{V}_j$ since the partitions $\mathcal{T}_j$ and $\mathcal{T}_j^*$ coincide on Ω_j by assumption. On the other hand, by Theorem 9 and 15 applied to $\{V_j^*\}$, we have for all $u \in V_J \subset V_J^*$

$$a(u,u) \approx (\|u\|_{A_{2,2}^1}^{***})^2 = \| \underbrace{\mathcal{Q}_0^* u}_{\in \tilde{V}_0} \|_2^2 + \sum_{j=1}^{J} 2^{2j} \| \underbrace{\mathcal{Q}_j^* u - \mathcal{Q}_{j-1}^* u}_{\in \tilde{V}_j} \|_2^2$$

which shows the assertion (cf. Theorem 16 and property **(A1)**).

The above result can also be interpreted from a different point of view (remember our admiration for basis functions rather than for partitions as the basic notion of our f.e. approximation schemes). The above nested refinement process is very close to a selection process of basis functions $N_{j,i}$ from a regular and quasi-uniform f.e. scheme $V_j = V(\mathcal{T}_j)$ satisfying all assumptions necessary for Theorem 15 which can be described as follows (note that the notation will be changed compared to the above one). Let $\{\Omega_j\}$ be any non-increasing sequence of subdomains in Ω which are the union of subdomains from the respective partitions $\mathcal{T}_{j-1}$, $j \ge 1$ ($\Omega_0 = \Omega$). Denote $\tilde{V}_j = \operatorname{span}\{N_{j,i} \,|\, \operatorname{supp} N_{j,i} \subset \Omega_j\}$, and introduce the final computational subspace by

$$V \equiv \sum_{j=0}^{J} \tilde{V}_j = \sum_{j=0}^{J} \sum_{\operatorname{supp} N_{j,i} \subset \Omega_j} V_{j,i}$$

This definition also explains what we call *nested selection* of basis functions.

It turns out (using the same argument, now with respect to the auxiliary sequence $\{V_j\}$ instead of $\{V_j^*\}$) that the additive Schwarz formulation for a variational problem **(VP)** on V corresponding to the above splittings (both BPX and MDS variants) possesses $O(1)$ condition number estimates which are independent of the choice of $\{\Omega_j\}$ and J. We subsume this in the following

Theorem 20. Subspaces of linear finite elements generated by nested refinement, or nested selection of basis functions as described above, allow for multilevel splittings that give rise to BPX resp. MDS preconditioners satisfying $O(1)$ condition number bounds which are independent of the number of refinement levels and of the particular refinement chosen.

This result can be generalized to other C^0 f.e. types that are used for second order elliptic problems. Thus, we have flexible and robust preconditioners for nested refinement and for nested selection of basis functions (by Theorem 18 we can run multiplicative Schwarz iterations on adaptively refined grids or subspaces as well). Their efficiency also depends on the growth of $\tilde{n}_j = \dim \tilde{V}_j$ since $\tilde{n} = \sum_{j=0}^J \tilde{n}_j$ might be somewhat larger than the dimension n of the computational subspace V. However, if the rules of nested refinement (or nested selection of basis functions) are followed then we have still $\tilde{n} = O(n)$, with a constant independent of $\{\Omega_j\}$. We will check this for nested selection in a triangular linear f.e. scheme. Indeed, the increase of dimension when adding $\tilde{V}_j$ to $\sum_{l=0}^{j-1} \tilde{V}_l$ exceeds the number of edges of $\mathcal{T}_{j-1}$ contained in the interior of Ω_j or in $\partial\Omega \cap \partial\Omega_j \backslash \Gamma$ since to each midpoint of such an edge there corresponds a basis function $N_{j,i} \in \tilde{V}_j$ which is linear independent of all basis functions used before on coarser grids. In turn, one third of the number of those edges is a true upper bound for $\tilde{n}_j$ since each basis function $N_{j,i}$ belonging to $\tilde{V}_j$ may be associated with exactly one of those edges, with a maximum of 3 basis functions per edge. Thus, we get

$$n \geq \tilde{n}_0 + \frac{1}{3} \sum_{j=1}^J \tilde{n}_j \geq \frac{1}{3}\tilde{n} \quad .$$

One might ask whether an $O(1)$ bound for the condition number of the additive Schwarz operator holds for arbitrary subsplitting of the BPX- resp. MDS-splitting, i.e., if V is given by an *arbitrary* sum of $V_{j,i}$. Simple examples (e.g. take only the subspaces of one fixed level j) show that this is not true. Thus, *subsplitting selection* is a more delicate task than the use of the refinement and clustering techniques described in 4.1. Only in the case of splittings **(D)** into direct sums of subspaces *any* subsplitting

$$\{\hat{V}; a\} = \sum_j \{\hat{V}_j; b_j\} \quad , \qquad \hat{V}_j \subset V_j \, ,$$

yields the same or even a better condition of the corresponding additive Schwarz operator. This is straightforward by Theorem 16. Note that some of the selected subspaces $\hat{V}_j$ may be trivial as well.

Another test case for modifying the basic splittings of Theorem 19 is the adaption to more general domain shapes (recall that our theory relies on a good resolution of Ω already by T_0, an assumption which is often in conflict with reality). A typical question which arises from the f.e. discretization of obstacle problems reads as follows. Let $\hat{\Omega} \subset \Omega$ be composed of triangles from the finest triangulation T_j used in a regular refinement process $V_0 \subset V_1 \subset \ldots \subset V_J$ as described above. By our theory, we have no problems with solving an $H_0^1(\Omega)$-elliptic variational problem with respect to V_J using the given multilevel hierarchy of subspaces V_j. In obstacle problems, however, one has to solve auxiliary problems with respect to

$$\hat{V}_J = \{g_J \in V_J \ : \ \operatorname{supp} g_J \subset \hat{\Omega}\} \,.$$

The main question is to find an iterative solution procedure (close to the multilevel splitting concept, if possible) which is robust with respect to the choice of $\hat{\Omega}$. A natural proposal [EHK, KY] is to examine the BPX- resp. MDS-splittings corresponding to

$$\hat{V}_0 \subset \hat{V}_1 \subset \ldots \subset \hat{V}_J \,, \quad \hat{V}_j = \hat{V}_J \cap V_j \,.$$

In the two-dimensional case, a satisfactory result has recently been obtained by Kornhuber and Yserentant [KY] who showed the robust estimate

$$\hat{\kappa}_J = O(J^2) \qquad \forall\, J \ \ \forall\, \hat{\Omega} \,.$$

Counterexamples [Os15] show that analogous results cannot be expected for three-dimensional problems, uniform $O(1)$ bounds need additional regularity assumptions on $\partial\hat{\Omega}$, see [EHK, KY, Os15]. We refer also to [HK, Ko] where different splittings are introduced and tested numerically.

4.2.3 Further developments and problems

The approach we have taken neglects some properties of a really applied elliptic problem. E.g., the size of the coefficients p, q of the differential operator has been ignored up to now, since we have substituted the energy norm $\sqrt{a(u, u)}$ by the problem independent norm $\|u\|_{H^1} \approx \|u\|_{A_{2,2}^1}$ to work out the precoditioning theory on the basis of the latter. We therefore may expect some detoriation of the rates of convergence of iterative methods when applying them to problems with strongly varying coefficients or to singularly perturbed problems which arise, for instance, in the method of lines for parabolic initial-boundary value problems. We have already mentioned the problems related to the effect of the shape of the domain and of the initial partition. A more problematic point is that the above theory requires *symmetric positive definite* elliptic problems and forms, see [Xu4, CX, BL2, CW, Wa1, Wa2] for the extensions of the theory to overcome this restriction.

The simplest model problem, the case of constant but different coefficients $p(x) = p_0 > 0$, $q(x) = q_0 \geq 0$, $x \in \Omega$, (for simplicity, let $\Gamma = \partial\Omega$) admits still a simple solution since under the assumptions of Theorem 15

$$a(u, u) \equiv p_0(\nabla u, \nabla u)_{L_2} + q_0(u, u)_{L_2} \approx \sum_{j=0}^{\infty}(p_0 2^{2j} + q_0)\|Q_j u - Q_{j-1} u\|_2^2$$

with constants in the norm equivalency that are *independent* of p_0, q_0. Q_j denotes the L_2 orthoprojection onto $V_j \cap H_0^1(\Omega)$. Thus, the splitting

$$\{V_J; a\} = \sum_{j=j_0}^{J} V_j = \sum_{j=j_0}^{J} \sum_{i} V_{j,i}$$

leads to condition numbers of the additive Schwarz operator uniformly bounded with respect to p_0, q_0, and J if the subspaces V_j resp. $V_{j,i}$ are equipped with the auxiliary forms $b_j(u, u) = (p_0 2^{2j} + q_0)(u, u)_{L_2}$. These can be replaced by the original form $a(\cdot, \cdot)$ on the one-dimensional subspaces in the last splitting. The number j_0 is determined by the condition $p_0 2^{2j_0} \approx q_0$ (if $p_0 > q_0$ then $j_0 = 0$, if $p_0 2^{2J} < q_0$ then there is no need in preconditioning, one may formally set $j_0 = J$).

Thus, by a simple strategy we can develop a coefficient-independent preconditioner of the same type as before (BPX or MDS) for the constant coefficient case, see [Ys4, BPS, Bm, Os12]. This question remains open in the general case. A proposal which works reasonably well on simple model problems and which localizes the above idea was made in [Os12] for the case $p(x) = 1$, $x \in \Omega$ and arbitrary $q(x) \geq 0$. Also, one may try Zhang's MDS which automatically introduces the size of the coefficients through diagonal scaling. However, there are by now no general modifications of low complexity of the existing multilevel preconditioners (or sufficient conditions for their application) that *guarantee* a coefficient-independent behaviour of the condition numbers, and thus of the rate of convergence of the preconditioned iterative methods under consideration. We want to mention this as an actual and difficult problem in the field. Generally, the problem of adapting multilevel preconditioning methods to a given problem can be put into the following abstract framework:

Given a variational problem (**VP**) *with s.p.d. form* $a(\cdot,\cdot)$ *in the computational Hilbert space* V, *find a splitting*

$$V = \sum_{j=0}^{J} V_j$$

and auxiliary s.p.d. forms $b_j(\cdot,\cdot) : V_j \times V_j \to \mathbf{R}$ *such that*

- *the condition number* $\mathrm{cond}(P)$ *of the additive Schwarz operator is small, and does not depend on certain parameters (coefficients etc.) of a (robustness),*

- *the overall arithmetical work (computational costs) to solve the subproblems and to add them together, i.e. to compute the action of the additive Schwarz operator* Pu *on an arbitrary* $u \in V$ *is* $O(n)$, *with* n *the dimension of* V.

If, for some reason, data structures are fixed then the degree of freedom in answering the question reduces to the optimal choice of the auxiliary forms $b_j(\cdot,\cdot)$. On the other hand, for certain problems one may even ask to choose first an adequate V of given approximation power, or to construct an optimal multilevel scale $\{V_j\}$ to properly resolve the continuous problem living in an infinite-dimensional function space. We do not expect general answers to these problems but feel that experimenting with the variable coefficient problem mentioned above might give some additional insight into how to finally formulate and solve such type of optimization problems. Note that the variable-coefficient problem might be too complex to expect final answers since in many cases hard non-symmetric problems of convection-diffusion type with dominating convection may be reduced to symmetric variable-coefficient problems. Last but not least, we want to mention some work on the so-called jumping coefficient case where the coefficients $p(x)$ resp. $q(x)$ are moderately changing inside each subdomain of $\mathcal{T}_0$ but may have large jump discontinuities along the subdomain boundaries (such assumptions are realistic for applied problems involving composite materials etc.). This case has been discussed in a number of papers in the domain decomposition circles [DW2, DW3, Ne5] but also in the context of multilevel preconditioning [Ys2, BY]. All these questions are not yet in final form, and further work is required. Some theoretical work by Bramble, Xu [BXu, Xu2] on the use of weighted L_2 projections in the variable coefficient problem shows the difficulties with our approach.

As usual, if elliptic solvers of high quality and robustness are available then parabolic initial-boundary value problems may be successfully attacked by the finite element method of lines, thus reducing the parabolic p.d.e. problem to a stiff system of ordinary differential equations. An alternative approach, the Rothe method, was recently taken up by Bornemann [Bm] to optimize adaptively both the time and space discretization. In both cases, the linear elliptic subproblem occuring in each implicit time step is equivalent to a Helmholtz-type problem where the above discussion applies [Ys4, Os12, Bm]. Anyhow, parabolic (and nonlinear elliptic) problems are fields where new ideas on preconditioning and fast solvers are welcome since the subroutine serving the elliptic problem will be called many times. Here, the coefficient-independence and robustness of preconditioning is highly desirable, too.

4.3   The biharmonic problem

In this short section, we comment on the results of [Os3, DOS, Zh2, Os8] concerning preconditioners for conforming C^1 f.e. discretizations of the first boundary value problem for the biharmonic problem:

$$\Delta^2 u = f \quad , \qquad u \in H_0^2(\Omega) \, .$$

This fourth order equation allows for very different symmetric variational formulations corresponding to different energy functionals used in plate theory. Other boundary conditions are possible, too. In order to apply the above theory, we have to work in subspaces of $H^2(\Omega)$ such that the underlying bilinear form is elliptic with respect to this Sobolev space.

The following element types have been discussed along the lines of subsection 3.4: Powell-Sabin elements [Os3], modified triangular and quadrilateral Clough-Tocher elements [Os3, DOS], C^1-quintics [DOS], the rectangular Bogner-Fox-Schmit element [Os8]. Some other conforming elements have been treated in [Zh2] by an analogous method. We refer also to an approach taken by Dörfler [Df1, Df2] which is close in spirit to the papers by Yserentant [Ys2, Ys3]. Figure 17 illustrates the structure of the local interpolation problem of some conforming finite element constructions for the biharmonic problem. Thus, we have a number of robust preconditioned multilevel solvers of asymptotically optimal efficiency (e.g. with condition numbers uniformly bounded with respect to the fine grid level, with a cheap V-cycle-like structure, and inherited stability with respect to adaptive refinement or basis function selection).

Note, however, that there are much more difficulties to overcome for higher order elliptic problems and C^1 elements. Even for some classical elements (Argyris or Bell triangles, standard and reduced triangular Clough-Tocher element) basic assumptions of our theory such as the monotonicity condition (M) are violated. Then we recommend the use of the above tricks: complicated elements may

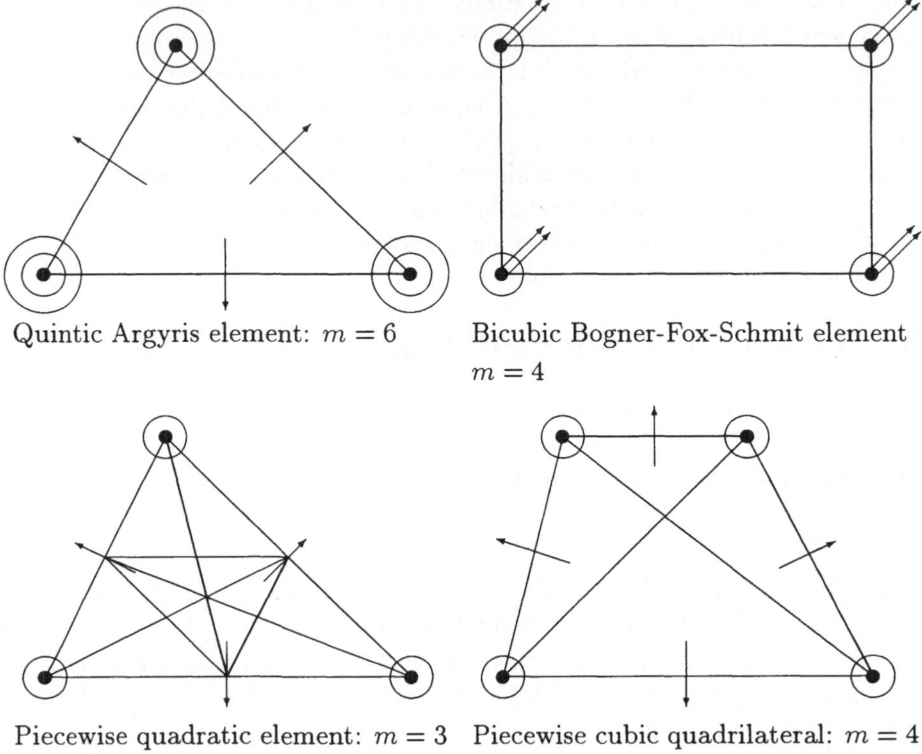

Quintic Argyris element: $m = 6$

Bicubic Bogner-Fox-Schmit element
$m = 4$

Piecewise quadratic element: $m = 3$ Piecewise cubic quadrilateral: $m = 4$
(Powell-Sabin split)

Figure 17. Examples of conforming C^1 elements

be treated by using preconditioners originally designed for low order elements (recall that for second order problems, the linear element preconditioner can be applied for this purpose). E.g., Powell-Sabin elements are good candidates for triangular partitions. Compare [Os16, Zh2].

It should be mentioned that the main bulk of engineering applications involving fourth order problems applies mixed formulations or nonconforming elements instead of the conforming element types considered in our approach. It is therefore an important problem to deal with these methods, and to develop an analogous theory of multilevel additive and multiplicative Schwarz methods for more realistic plate and shell models. Compare 4.6.

4.4 Domain decomposition and boundary element methods

We do not attempt to cover these fields of recent intensive research activities and developments to any depth. We only want to show by *examples* that the multilevel splittings of Theorem 19, in full analogy to what was explained in 3.4.4, yield almost automatically well-behaved multilevel splittings for variational problems on inner resp. boundary manifolds. See [Os12, Os8, Os10]. Such problems naturally arise when so-called nonoverlapping domain decomposition methods resp. boundary element methods are considered. Other aspects, e.g., domain decomposition with overlapping regions, are only briefly touched on.

Let us consider the simplest situation of linear finite element subspaces $V_j = V(T_j)$ produced by regular dyadic refinement of a polyhedral domain in $\mathbf{R}^d$. To avoid technical discussions, let the underlying problem be $H^1(\Omega)$, i.e. $\Gamma = \emptyset$. We fix the computational subspace to be $V = V_J$, i.e. the number of levels is J. We also fix a second integer $j_0 \in [0, J)$ which represents the level number of the coarse grid. The open subdomains of T_{j_0} are denoted by Ω_k, the union of their boundaries will be denoted by γ, see Figure 18 a). One may also group a few subdomains together into one Ω_k but their number should be bounded by an absolute constant to keep the size of Ω_k comparable with the mesh-size of T_{j_0}. The nodal basis coefficient vector corresponding to the solution of **(VP)** splits into

$$x \equiv (x_I, x_\gamma), \quad x_I \equiv (x_{\Omega_1}, x_{\Omega_2}, \dots)$$

where x_{Ω_k} resp. x_γ represent the unknowns associated with nodal points interior to Ω_k resp. to γ. Accordingly, the linear system (nodal basis discretization) corresponding to **(VP)** takes the form

$$\begin{pmatrix} A_{II} & A_{I\gamma} \\ A_{\gamma I} & A_{\gamma\gamma} \end{pmatrix} \begin{pmatrix} x_I \\ x_\gamma \end{pmatrix} = \begin{pmatrix} b_I \\ b_\gamma \end{pmatrix}.$$

Observe that A_{II} is block–diagonal, with the blocks corresponding to elliptic problems on the subregions Ω_i which are of smaller size and can be solved in parallel. We will assume that we possess an exact solver for these subproblems, i.e. for computing $A_{II}^{-1} b_I$. Then the system may be reduced formally to a smaller one

$$S_\gamma x_\gamma \equiv (A_{\gamma\gamma} - A_{\gamma I} A_{II}^{-1} A_{I\gamma}) x_\gamma = b_\gamma - A_{\gamma I} A_{II}^{-1} b_I.$$

The Schur complement matrix S_γ is still badly conditioned if $J \to \infty$, and we look for a preconditioner.

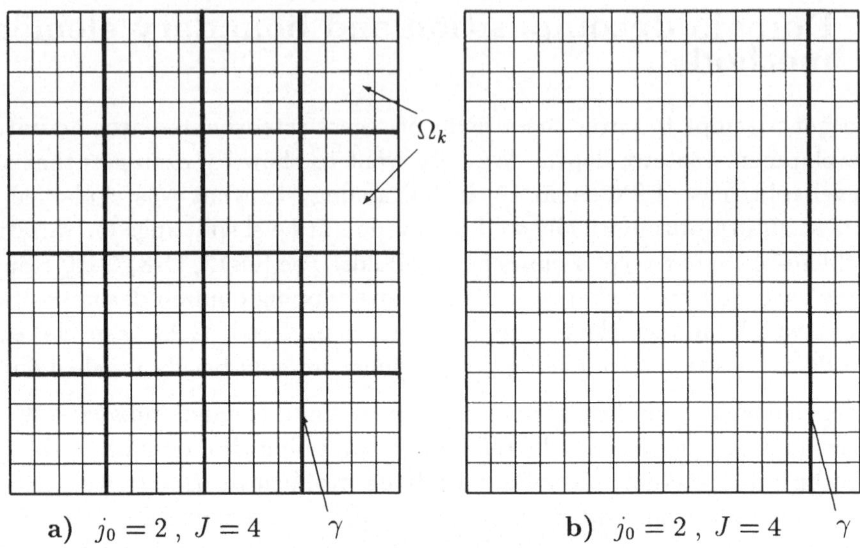

Figure 18. Nonoverlapping domain decomposition

The derivation of preconditioners which are analogs of the HB, BPX, resp. MDS-methods has been given in [Os12] (see also [SW, TCK]) by using the variational characterization of the Schur complement: if we denote by $V_{j,\gamma}$ the trace spaces of V_j onto γ which turn out to be also linear finite element spaces with respect to the partitions induced by $\mathcal{T}_j$ on γ then

$$((S_\gamma x_\gamma, x_\gamma)) \equiv s_\gamma(u_\gamma, u_\gamma) = \inf_{u \in V_J : u|_\gamma = u_\gamma} a(u, u) , \quad u_\gamma \in V_\gamma \equiv V_{J,\gamma} .$$

Indeed, denoting $x_{\gamma,h} = -A_{II}^{-1} A_{I\gamma} x_\gamma$, one has the identity

$$\begin{aligned} a(u, u) &= ((A_{II} x_I, x_I)) + 2((A_{I\gamma} x_\gamma, x_I)) + ((A_{\gamma\gamma} x_\gamma, \gamma)) \\ &= ((A_{II}(x_I - x_{\gamma,h}), x_I - x_{\gamma,h})) + ((S_\gamma x_\gamma, x_\gamma)) \end{aligned}$$

which gives the assertion, the minimum is attained if $x_I = x_{\gamma,h}$. The corresponding f.e. function $u_{\gamma,h}$ is called harmonic (with respect to the elliptic form a) extension of $u_\gamma \in V_\gamma$ and plays an important role in all kind of non-overlapping domain decomposition algorithms. Finally, $((\cdot, \cdot))$ denotes the scalar product in the respective Euclidean space of real vectors. Comparing this with our definition and construction of trace spaces, cf. 3.4.4, we get good splittings from Theorem 13 (specialized to the particular case needed here). See [Os12] for the details and a self-contained discussion.

Here we go an alternative way [Os15] to obtain the same result. We start with our basic splittings from Theorem 19 and use clustering arguments. Indeed, let us take the BPX-splitting

$$V = V_J = \sum_{j=0}^{J} \sum_i V_{j,i} \,, \quad b_{j,i}(u,v) = 2^{2j}(u,v)_{L_2} \,,$$

and group the one-dimensional subspaces as follows:

$$V = \underbrace{\sum_{j=0}^{j_0} \sum_i V_{j,i}}_{V_{j_0}} + \sum_k \underbrace{\sum_{j=j_0+1}^{J} \sum_{i\,:\,\mathrm{supp}\,N_{j,i} \subset \Omega_k} V_{j,i}}_{V_{\Omega_k}} + \underbrace{\sum_{j=j_0+1}^{J} \sum_{i\,:\,\mathrm{supp}\,N_{j,i} \cap \gamma \neq \emptyset} V_{j,i}}_{\tilde V_\gamma} \,.$$

The first group represents a splitting of V_{j_0} (equipped with $a(\cdot,\cdot)$ as the basic bilinear form). The second term contains the splittings for the subspaces $V_{\Omega_k} = \{v \in V \,|\, \mathrm{supp}\, v \subset \Omega_k\}$ corresponding to the subregions and also equipped with the scalar product $a(\cdot,\cdot)$. By Theorem 19 (and the clustering argument) we can replace the splittings by the respective subspaces without destroying the behavior of the condition numbers of the additive Schwarz operator. The latter is equivalent to the norm equivalence

$$a(u,u) \;\approx\; \inf_{u = u_{j_0} + \sum u_{\Omega_k} + \sum \sum u_{j,i}} \Big(a(u_{j_0}, u_{j_0}) + \sum_k a(u_{\Omega_k}, u_{\Omega_k})$$
$$+ \sum_{j_0+1} \sum_{P_{j,i} \in \gamma} 2^{2j} \|u_{j,i}\|_{L_2}^2 \Big)$$

which holds for all $u \in V$. Now we use the variational definition of $s(\cdot,\cdot)$ substituting the above norm equivalence: For arbitrary $u_\gamma \in V_\gamma$

$$s(u_\gamma, u_\gamma) \;\approx\; \inf_{u \in V\,:\,u|_\gamma = u_\gamma} \; \inf_{u = u_{j_0} + \sum u_{\Omega_k} + \sum \sum u_{j,i}} \Big(a(u_{j_0}, u_{j_0})$$
$$+ \sum_k a(u_{\Omega_k}, u_{\Omega_k}) + \sum_{j_0+1}^{J} \sum_{P_{j,i} \in \gamma} 2^{2j} \|u_{j,i}\|_{L_2}^2 \Big)$$
$$= \inf_{u_\gamma = \left(u_{j_0} + \sum \sum u_{j,i} \right)|_\gamma} a(u_{j_0}, u_{j_0}) + \sum_{j_0+1}^{J} \sum_{P_{j,i} \in \gamma} 2^{2j} \|u_{j,i}\|_{L_2}^2 \,.$$

We now see the additive Schwarz splitting of V_γ suitable for preconditioning the above Schur complement problem:

$$V_\gamma = V_{j_0,\gamma} + \sum_{j_0+1}^{J} \sum_{P_{j,i} \in \gamma} V_{j,i,\gamma}$$

where $V_{j,i,\gamma} = V_{j,i}|_\gamma = \text{span } N_{j,i}|_\gamma$ is equipped with the scalar product

$$b_{j,i,\gamma}(u_{j,i,\gamma}, v_{j,i,\gamma}) = 2^j (u_{j,i,\gamma}, v_{j,i,\gamma})_{L_2(\gamma)} \approx 2^{2j}(u_{j,i}, v_{j,i})_{L_2}$$

and $V_{j_0,\gamma}$ with $b_{j_0,\gamma}(u_{j_0,\gamma}, v_{j_0,\gamma}) = a(u_{j_0}, v_{j_0})$ (note that any function $u_{j_0,\gamma} \in V_{j_0,\gamma}$ uniquely extends to a function $u_{j_0} \in V_{j_0}$).

The resulting preconditioner yields an $O(1)$ bound for the condition number of the preconditioned Schur complement problem. It has been tested practically in [TCK] where it was derived in a different way. If we start (in the two-dimensional setting) with the HB-splitting then we get a method which was examined in [SW] and has an asymptotics of $O((J - j_0)^2)$ for the corresponding condition numbers.

Following some tradition in the domain decomposition circles one may group together the one-dimensional subspaces with respect to vertices, edges, faces etc. of γ resp. the coarse level partition $\mathcal{T}_{j_0}$. This would lead to some better parallelism also in the preconditioning step of a potential code using the above Schur complement technique (the multiplication by S_γ involves the solution of the independent subproblems on the Ω_k, a task which can be carried out in parallel). Compare also the wirebasket preconditioners [Sm, DW3].

Analogous derivations are possible, in principle, with any f.e. scheme which allows for analogs of Theorem 19. E.g., in the case of the biharmonic equation we can derive, in complete analogy to the above considerations, a Schur complement preconditioner using cubic Hermite splines on γ from the results on the Bogner-Fox-Schmit rectangle obtained in [Os8].

The most striking restriction of the approach is the assumption that the coarse level partition into the subdomains Ω_k coincides with some $\mathcal{T}_{j_0}$. It is not very hard to prove the optimality (with respect to condition number behavior) of an analogous preconditioner if $\{\Omega_k\}$ is *any* quasi-uniform partition of Ω into subdomains compatible with the f.e. type, and the fine level partition is a refinement of this coarse level partition. But one may try more complicated situations as well. As a simplified model case, take the linear finite element case and consider a subdivision of a square into two (non-equal) rectangles Ω_1 and Ω_2 with the common boundary γ, without loss of generality we suppose compatibility with some sequence of rectangular partitions $\{\mathcal{T}_j\}$ in the following sense: the two rectangles are composed of rectangles from $\mathcal{T}_{j_0}$ as shown in Figure 18 b). The final level number is $J \geq j_0$.

We consider once again our basic BPX-splitting from Theorem 19. Then, using the same notation as above, for the Schur complement problem on γ we get

$$
\begin{aligned}
s(u_\gamma, u_\gamma) &\approx \inf_{u \in V:\, u|_\gamma = u_\gamma} \;\; \inf_{u = \sum_{j=0}^{J} \sum_i u_{j,i}} \left(\sum_{j=0}^{J} \sum_i 2^{2j} \|u_{j,i}\|_{L_2}^2 \right) \\
&= \inf_{u_\gamma = \sum_{j=0}^{J} \sum_i u_{j,i}|_\gamma} \left(\sum_{j=0}^{J} \sum_{N_{j,i}|_\gamma \neq 0} 2^{2j} \|u_{j,i}\|_{L_2}^2 \right) \\
&\approx \inf_{u_\gamma = \sum_j \sum_i u_{j,i,\gamma}} \left(\sum_{j=0}^{J} \sum_i 2^{j} \|u_{j,i,\gamma}\|_{L_2(\gamma)}^2 \right) .
\end{aligned}
$$

The last expression corresponds to the one-dimensional BPX-splitting of V_γ and is equivalent to the square of the $H^{1/2}(\gamma)$ norm, compare the definition of trace spaces and Theorem 13. The last step in the estimation follows if one looks at the traces of the two-dimensional basis functions $N_{j,i}$ to γ. In the case of Dirichlet data, the result of this comparison is different: it yields a correct scaling factor $2^{\max(j,jo)}$ instead of 2^j.

We do not go into further detail. What should be clear (by the above examples, and by the whole theory of trace spaces behind) is that we can get splittings for f.e. subspaces and variational problems on boundary manifolds by starting with the splittings of BPX type which are sufficiently fine, allow for simple structured multilevel algorithms, and can be used for deriving a lot of different variants by clustering etc.. Local refinement is also obvious, compare 4.2.2 for the basic splittings.

The approach also applies to overlapping domain decomposition schemes. One can give alternative derivations of adequate coarse grid problems depending on the amount of overlap. This does not mean that the way via Theorem 19 is the only resp. the optimal one. We refer the interested reader to the many, many research papers, especially the series of domain decomposition conference proceedings containing different approaches and far more material.

We briefly comment on the question of whether the above methods are useful for boundary element methods as well. There one has typically symmetric (or nonsymmetric) elliptic problems with respect to Sobolev spaces on $\partial\Omega$. The interesting cases where preconditioning is necessary are smoothness exponents $s = 1/2$ resp. $s = -1/2$. As was mentioned in 3.4.4, the space $H^{1/2}(\partial\Omega)$ is a boundary trace space of $H^1(\Omega)$, and possesses a description by approximation spaces (Theorem 13). This yields analogs of Theorem 19, and by our theory also good additive and multiplicative Schwarz methods for symmetric $H^{1/2}$ elliptic problems. The case of $s = -1/2$ was discussed in [BL1, Os10, Os15] (an example of practical interest is the single layer potential integral equation of the first kind). It requires additional duality arguments since, till now, we are able to obtain splittings only for Sobolev spaces of positive smoothness. We sketch one result for this case. Suppose that $a(\cdot,\cdot)$ is a symmetric $H^{-1/2}(\partial\Omega)$-elliptic bilinear form. Let $\{W_j\}$ be an appropriate sequence of finite element spaces with respect to the polyhedral boundary manifold $\partial\Omega$ (one may think about Lagrange C^0 element spaces induced by the corresponding construction for Ω). Suppose (as may be expected from the above theory) that

$$\|u\|^2_{H^{1/2}(\partial\Omega)} \approx \inf_{w_j \in W_j \,:\, u = \sum_j w_j} \sum_{j=0}^{\infty} 2^j \|w_j\|^2_{L_2(\partial\Omega)} \quad \forall\, u \in H^{1/2}(\partial\Omega)\,,$$

with the corresponding corollaries if $u \in W \equiv W_J$ for some arbitrarily fixed J. To produce subspaces and stable splittings with respect to $H^{-1/2}(\partial\Omega)$ one needs *isomorphisms* $S : W \to S(W)$, i.e. satisfying

$$\|Sw\|_{H^{-1/2}(\partial\Omega)} \approx \|w\|_{H^{1/2}(\partial\Omega)} \quad \forall\, w \in W\,.$$

A preliminary splitting suitable for the $H^{-1/2}$ problem is then given by

$$S(W) = \sum_{j=0}^{J} S(W_j)\,, \quad b_j(Sw_j, Sv_j) = 2^j(w_j, v_j)_{L_2(\partial\Omega)}\,,$$

with w_j, v_j denoting arbitrary elements in W_j, $j = 0,\dots,J$. Since the corresponding Schwarz algorithms would make permanent use of S, the action of S should be maximally explicit and local.

If the boundary manifold is one-dimensional (i.e. if $\Omega \subset \mathbf{R}^2$) there are obvious candidates for this task: simple differentiation will do the job, cf. [Os10]. However, in higher dimensions this is by no means obvious. Prewavelets with respect to the multiresolution analysis

$$W_0 \subset W_1 \subset \dots \subset W_J(= W) \subset \dots \subset L_2(\partial\Omega)$$

might be helpful in this respect. Indeed, let $\tilde{W}_j = W_j \ominus W_{j-1}$ resp. W_0 for $j > 0$ resp. $j = 0$ (the orthogonal complements taken with respect to the L_2 scalar product on $\partial\Omega$), and consider locally supported prewavelet bases $\{\tilde{\psi}_{j,i}\}$ in $\tilde{W}_j$. As a consequence of the fact that these basis sets form Riesz bases, uniformly in j, and from the mutual orthogonality of the $\tilde{W}_j$, we obtain uniformly in J

$$\|u\|^2_{H^{1/2}(\partial\Omega)} \approx \sum_{j=0}^{J}\sum_i 2^j \tilde{c}_{j,i}^2 \quad \forall\, u \equiv \sum_{j=0}^{J}\sum_i \tilde{c}_{j,i}\tilde{\psi}_{j,i}$$

(here, the prewavelet functions have been scaled to 1 in the L_2 norm). We will now show that the mapping $S : W \to W$ given by

$$Su = \sum_{j=0}^{J} \sum_{i} 2^j \tilde{c}_{j,i} \tilde{\psi}_{j,i}$$

fits the above requirements. By definition of the $H^{-1/2}$ norm,

$$\|Su\|_{H^{-1/2}(\partial\Omega)} = \sup_{0 \neq v \in H^{1/2}(\partial\Omega)} \frac{(Su, v)_{L_2(\partial\Omega)}}{\|v\|_{H^{1/2}(\partial\Omega)}}$$

$$\leq C \max_{0 \neq v \in W} \frac{\sum_{j=0}^{J} 2^j (\tilde{u}_j, \tilde{v}_j)_{L_2}}{\left(\sum_{j=0}^{J} 2^j \|\tilde{v}_j\|_{L_2}^2\right)^{1/2}} \leq C \left(\sum_{j=0}^{J} 2^j \|\tilde{u}_j\|_{L_2}^2\right)^{1/2}$$

$$\leq C\|u\|_{H^{1/2}(\partial\Omega)}$$

where $\tilde{u}_j$ resp. $\tilde{v}_j$ denote the L_2-orthoprojections of u resp. v onto $\tilde{W}_j$ and the above norm description in $H^{1/2}$ has been used several times. On the other hand, by analogous considerations

$$\|Su\|_{H^{-1/2}(\partial\Omega)} \geq \frac{(Su, u)_{L_2(\partial\Omega)}}{\|u\|_{H^{1/2}(\partial\Omega)}} \geq C\|u\|_{H^{-1/2}(\partial\Omega)} \ .$$

We do not know a reference for prewavelet constructions on general polyhedral manifolds but believe that this question can be settled (this is the analogon of the problem formulated at the beginning of subsection 4.2, compare [KO]). It is interesting to look at the resulting additive Schwarz operator. As before, for computational reasons we refine the splitting to a BPX-like splitting into one-dimensional subspaces $\tilde{W}_{j,i}$ corresponding to single prewavelet functions $\tilde{\psi}_{j,i}$:

$$S(W) = \sum_{j=0}^{J} \sum_{i} S(\tilde{W}_{j,i})$$

with the forms $\tilde{b}_{j,i}$ being the restrictions of the b_j. Since $S(\tilde{\psi}_{j,i}) = 2^j \tilde{\psi}_{j,i}$ and

$$b_j(S\tilde{\psi}_{j,i}, S\tilde{\psi}_{j,i}) = 2^j (\tilde{\psi}_{j,i}, \tilde{\psi}_{j,i})_{L_2(\partial\Omega)} = 2^j \ ,$$

due to the normalization assumption on the wavelets, we obtain the following simple structure of the V-cycle corresponding to the action of the additive Schwarz operator on an arbitrary function $u \in S(W) = W$:

$$Pu = \sum_{j=0}^{J} \sum_{i} \frac{a(u, S\tilde{\psi}_{j,i})}{b_j(S\tilde{\psi}_{j,i}, S\tilde{\psi}_{j,i})} S\tilde{\psi}_{j,i} = \sum_{j=0}^{J} \sum_{i} 2^j a(u, \tilde{\psi}_{j,i}) \tilde{\psi}_{j,i} \ .$$

It is up to the potential user to work now with the traditional nodal basis BEM discretization and to use the pyramid algorithm only in the preconditioning step, or to directly discretize the $H^{-1/2}$ problem with respect to the prewavelet basis. For boundary integral/element applications, (pre)wavelets have some additional advantages which make them especially attractive in this case: for large classes of kernels, they automatically lead to a better sparsity, in the sense that the decay of the coefficients away from the diagonal in the discretized linear system is much more considerable compared with usual nodal basis functions. Details are explained in papers by the Beylkin/Coifman/Rohlin group, e.g., [BC1, BC2], we refer also to [DP1, DP2]. Compare the alternative approaches by Brandt, Lubich [BLu], Hackbusch/Novak [HN] where fast matrix-vector multiplications with respect to the usual nodal basis discretization are proposed. Anyhow, we see that even problems involving negative Sobolev exponents can be approached. See also [BL1] for a multiplicative Schwarz method in the model problem of a single layer potential equation on a unit square in $\mathbf{R}^2$.

4.5   Sparse grids

This short section is devoted to a discussion of the sparse grid discretization technique introduced by Zenger [Ze] and his group within our framework. We will be brief, for details we refer to [GO1] (some additional material is, however, given). Let $\Omega \equiv [0,1]^d$ be the d-dimensional cube ($d \geq 2$). For given $\mathbf{k} \in \mathbf{Z}_+^d$, $\mathbf{k} \equiv (k_1, \ldots, k_d)$, let $\mathcal{R}_{\mathbf{k}}$ be the tensor-product partition with uniform stepsize 2^{-k_j} into the j-th coordinate direction, $j = 1, \ldots, d$. By $V_{\mathbf{k}}$ we denote the spaces of multilinear finite element functions with respect to $\mathcal{R}_{\mathbf{k}}$, $\mathbf{k} \in \mathbf{Z}_+^d$. If homogeneous boundary conditions are required, we have to consider $V_{\mathbf{k},0} = V_{\mathbf{k}} \cap H_0^1(\Omega)$ where $\mathbf{k} \geq \mathbf{1} \equiv (1, \ldots, 1)$ is a natural restriction. Note that higher order C^0-elements, C^1-element spaces for treating fourth order problems, and slightly more complicated boundary conditions may be considered as well, we restrict our attention to the simplest above-mentioned case. The spaces $V_{\mathbf{k}}$ satisfy the following monotonicity property

$$V_{\mathbf{k}} \subset V_{\mathbf{m}} \quad , \qquad \mathbf{k} \leq \mathbf{m}.$$

Now we introduce, for $k \in \mathbf{Z}_+$, the full grid spaces

$$\hat{V}_k = V_{(k,\ldots,k)} \quad (k > 0)$$

and the sparse grid spaces

$$\tilde{V}_k = \sum_{|\mathbf{k}|_1 \leq k} V_{\mathbf{k}} \quad (k \geq d) .$$

(here and in the following we denote $|\mathbf{k}|_1 = \sum_j k_j$ and $|\mathbf{k}|_\infty = \max_j k_j$ whenever

$\mathbf{k} \in \mathbf{Z}_+^d$). As is pointed out in a series of papers by the Munich group, compared with the full grid spaces, the sparse grid spaces $\tilde{V}_k$ enjoy almost the same approximation properties under mild additional smoothness assumptions on the functions (solutions of second order elliptic problems) to be approximated but their dimension is considerably smaller. Especially for $d = 3$ the difference is striking [Bg]. Note that a proper study of the approximation power of these new subspaces has still to be carried out, it requires the theory of spaces with dominating mixed derivatives which were briefly mentioned in 3.1, and is more complicated than in the full grid case. Nevertheless, we are able to derive splittings for $\tilde{V}_k$ without knowing too much.

The sequences of full grid spaces fit the above framework. E.g., Theorem 19 resp. 15 are applicable. To get some consequences out of the norm equivalencies known for the full grid spaces, introduce the following notation. Let W_j, $j \geq 0$ the (pre)wavelet spaces corresponding to the *one-dimensional* multiresolution analysis formed by linear elements (= linear splines) restricted to $[0, 1]$, and consider their d-dimensional counterparts $W_{\mathbf{j}}^{(d)}$ obtained as tensor products from the one-dimensional copies $W_{j_1}, \ldots, W_{j_d}$ corresponding to the coordinate directions $x_1, \ldots, x_d$. Obviously, we have the following L_2-orthogonal decompositions into *direct* sums of subspaces:

$$\hat{W}_k \equiv \hat{V}_k \ominus \hat{V}_{k-1} = \bigoplus_{|\mathbf{j}|_\infty = k} W_{\mathbf{j}}^{(d)} \quad ,$$

$$\tilde{W}_k \equiv \tilde{V}_J \cap \hat{W}_k = \bigoplus_{|\mathbf{j}|_\infty = k, |\mathbf{j}|_1 \leq J} W_{\mathbf{j}}^{(d)} \quad , \quad k \leq J.$$

Due to these equalities and one assertion of Theorem 15 (involving the ortho-projections Q_j) we see that the decomposition

$$\tilde{V}_J = \sum_{k=0}^J \tilde{W}_k = \sum_{k=0}^J \sum_{|\mathbf{j}|_\infty = k, |\mathbf{j}|_1 \leq J} W_{\mathbf{j}}^{(d)}$$

is L_2-orthogonal and H^1 stable (with stability constants bounded independently of J) if the L_2 scalar product scaled by the factor 2^{2k} is used as the auxiliary bilinear form on $\tilde{W}_k$ and its subspaces.

One could derive a practical method directly from this observation: local and L_2-stable bases serving for a final BPX splitting into one-dimensional subspaces of $W_{\mathbf{j}}^{(d)}$ can be formed by tensor products of *one-dimensional* linear prewavelets. More general, any one-dimensional prewavelet or wavelet scheme leads via simple tensor-product constructions to a good splitting of the corresponding *sparse wavelet space.* It seems that, for some historical reason, people from the wavelet circles have ignored (till now!) this trivial possibility to write papers and to obtain useful results.

Another, more traditional way was followed in [GO1] (the proof was slightly different): Using the relationships

$$\tilde{W}_k \subset \sum_{|\mathbf{j}|_\infty = k, |\mathbf{j}|_1 = J} V_{\mathbf{j}} \subset \hat{V}_J \cap \hat{V}_k \,, \quad kd > J$$

(note that the number of subspaces $V_{\mathbf{j}}$ involved for each k does not exceed CJ^{d-2}) and $\tilde{W}_k \subset \hat{V}_k \subset \tilde{V}_J$ for $kd \leq J$, one obtains a new splitting

$$\tilde{V}_J = \sum_{k \,:\, kd \leq J} \hat{V}_k + \sum_{k \,:\, kd > J} \sum_{|\mathbf{j}|_\infty = k, |\mathbf{j}|_1 = J} V_{\mathbf{j}} \,,$$

with an upper estimate of the condition number of the corresponding BPX preconditioner given qualitatively by $O(J^{d-2})$. Now, instead of tensor products of prewavelets, tensor products of univariate nodal basis functions can be used which leads to smaller computational masks but to slightly more functions having to be considered. Note that the paper [GO1] gives also an analysis of the hierarchical basis approach to sparse grid spaces which is much worse than the above results, both in theory and in practice.

There are completely different preconditioners (e.g., the combination technique) and multigrid-like methods [Gr1, Gr2] for the sparse grid case which are promising but require further analysis. We hope to have encouraged the interested reader to contribute to the sparse grid discretization technique which is attractive for higher-dimensional problems.

We conclude with a remark: sparse grid techniques (or better: methods to analyze them) may be useful also when significant dependencies in coordinate directions are present. This is the case for anisotropic problems like

$$\epsilon \frac{\partial^2 u}{\partial x^2} + \frac{\partial^2 u}{\partial y^2} = f \quad , \quad (x,y) \in \Omega \subset \mathbf{R}^2 \,,$$

for problems on stretched domains and/or partitions with stretched rectangles etc.. I feel that after all we are now experienced enough to believe in this possibility (see [GO1] for some observations concerning computations with the sparse BPX method for the model anisotropic problem, and [Ha3, Ha4, RWZ] for multigrid approaches which are close to the Schwarz setting).

4.6 Nonconforming and mixed methods

4.6.1 Splittings for nonconforming methods

We recall that the fictitious space lemma (Theorem 17) allows the construction of preconditioners also in cases if the auxiliary problems, one wants or has to use, do not allow for an immediate interpretation in terms of subproblems in subspaces of the computational space. This is typically the case if nonconforming methods are considered. We comment on some multilevel schemes which were derived over the last few years, without relying too much on this theoretical construction. Figure 19 shows the local interpolation problems for some nonconforming methods discussed in this short subsection.

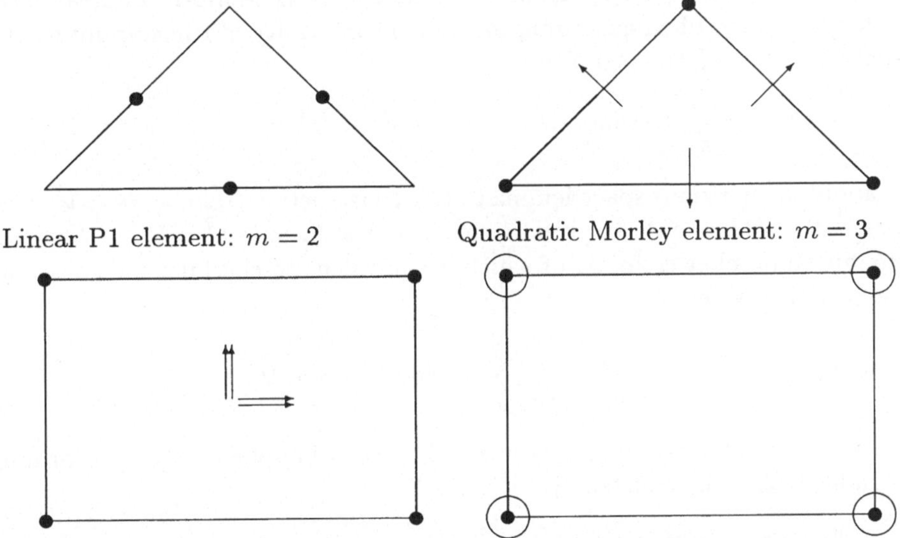

Figure 19. Typical nonconforming elements

Many papers have been written on the simplest scalar P1 (Crouzeix-Raviart) element [Xu2, Os7, Os11, Df1, Df2, VW2]. The following derivation in the spirit of the fictitious space lemma is based on a simple idea that may be used for other nonconforming multilevel schemes as well. Denote by $\mathcal{V}_j$ the sequence of nonconforming P1 elements, by V_j the sequence of usual conforming linear elements, both constructed with respect to a dyadically refined sequence of triangulations of $\Omega \subset \mathbf{R}^d$. Recall that the degrees of freedom of the P1 elements are supported at the midpoints M of the $(d-1)$-dimensional faces of the elements while functions from V_j are uniquely given by their values at the vertices. Fix $\mathcal{V} \equiv \mathcal{V}_J$ for the computational subspace, and define the mapping

$$R : \mathcal{V}_0 \times \mathcal{V}_1 \times \ldots \mathcal{V}_J \to \mathcal{V} ,$$

by $R(u_0, \ldots, u_j) = \sum_{j=0}^{J} R_j u_j$, with $R_J = \mathrm{Id}_{\mathcal{V}_J}$ and $R_j = I_J I_{J,j} \mathcal{Q}_j$ where $I_{J,j}$ resp. I_J denote the natural imbeddings from V_j into V_J resp. V_J into $\mathcal{V}_J$ and $\mathcal{Q}_j : \mathcal{V}_j \to V_j$ is the restriction of the quasi-interpolant operator from subsection 2.1.1 onto V_j. As before, take

$$b_j(u_j, v_j) = 2^{2j}(u_j, v_j)_{L_2} , \quad u_j, v_j \in \mathcal{V}_j , \; j \ge 0 .$$

The form on $\mathcal{V}$ which comes from a symmetric $H^1(\Omega)$-elliptic bilinear form $a(\cdot, \cdot)$ has to be modified according to the nonconformity of the computational subspace $\mathcal{V}$ following the usual rules:

$$a_J(u, v) = \sum_{K \in \mathcal{T}_J} a(u|_K, v|_K) \quad \forall\, u, v \in \mathcal{V} .$$

Now, apply the fictitious space lemma to this construction (taking $H_0 = \mathcal{V}$, $H = \mathcal{V}_0 \times \ldots \times \mathcal{V}_J$, and so on). It gives rise to a preconditioner $C = \sum_{j=0}^{J} R_j B_j^{-1} R_j^*$ the condition number estimate of which is equivalent to checking the constants of the norm equivalence

$$a_J(u, u) \approx \inf_{u_j \in \mathcal{V}_j \,:\, u = \sum_{j=0}^{J} R_j u_j} \sum_{j=0}^{J} 2^{2j} \|u_j\|_2^2 \quad \forall\, u \in \mathcal{V} .$$

But the latter is almost immediate from Theorem 19 applied to the conforming case. Indeed, since by construction

$$\|v_j\|_2 \approx \inf_{u_j \in \mathcal{V}_j \,:\, v_j = \mathcal{Q}_j u_j} \|u_j\|_2 \quad \forall\, v_j \in V_j , \; j = 0, \ldots, J,$$

we have

$$\begin{aligned}
a(v, v) &\approx \inf_{v_j \in V_j \,:\, v = \sum_{j=0}^{J} v_j} \sum_{j=0}^{J} 2^{2j} \|v_j\|_2^2 \\
&\approx \inf_{u_j \in \mathcal{V}_j \,:\, v = \sum_{j=0}^{J} I_{J,j} \mathcal{Q}_j u_j} b_j(u_j, u_j) \qquad \forall\, v \in V_J .
\end{aligned}$$

Thus, the desired norm equivalence is reduced to proving

$$a_j(u, u) \approx \inf_{\hat{u} \in \mathcal{V}, v \in V_J \,:\, u = \hat{u} + v} \{ 2^{2J} \|\hat{u}\|_2^2 + a(v, v) \}$$

which is left to the reader (see [Os11]) as an exercise. The double occurence of

$\mathcal{V}_J$ may be deleted by using the clustering argument (alternatively, one may use $\mathcal{V} = \mathcal{V} + \mathcal{V}_{J-1}$ as intermediate splitting, see [VW2]).

Note, that in order to make the scheme attractive for computations one would prefer to switch to a BPX-like formulation by discretizing the L_2 forms with respect to the nodal bases in $\mathcal{V}_j$. This also makes explicit the adjoints $R^*_{j,i}$ that are required during the computations. One may look for further improvements, e.g. one should carefully examine the constructions with respect to decreasing computational efforts, e.g., by keeping the masks for the level-to-level mappings as small as possible.

We give a list of results on nonconforming multilevel schemes which have been derived along the lines of the above approach: Wilson rectangles were treated in [Os4], another decomposition (of hierarchical basis type) for the nonconforming P1 element was discussed in [Os7], [Df1, Df2] contain a more general construction which is, however, not yet in final shape. The Adini element for plate bending has recently been considered in [Os8] (here, the Bogner-Fox-Schmit rectangle plays the role of the conforming counterpart). We also mention our paper [Os16] where some more examples are given, illustrating the simple strategy of switching to a reference multilevel method to avoid the difficulties with nonnested nonconforming f.e. spaces. To complete the picture, one may also look at papers by Bramble, Pasciak, Wang, Xu et.al. following the ideas of [BX2, BW1, BW2] (for instance, [Hn] contains some results on multigrid schemes for the Morley discretization), and at a series of papers by Brenner [Bn1, Bn2, Bn3, Bn4].

4.6.2 Mixed finite element methods

We conclude our rush through the areas of possible applications of the basic norm equivalencies of Theorem 15 and 19 by touching on mixed finite element constructions. The increasing popularity of mixed methods has both physical and computational background. We give an illustrative example. Consider a standard elliptic problem with homogeneous Neumann type boundary conditions

$$-\nabla(p\nabla u) + qu = f \quad \text{in } \Omega \subset \mathbf{R}^d , \qquad \frac{\partial u}{\partial n} = 0 \quad \text{on } \partial\Omega .$$

The corresponding variational problem **(VP)** defined on the Sobolev space $H^1(\Omega)$ uses the bilinear form

$$a(u,v) \equiv (p\nabla u, \nabla v)_{(L_2(\Omega))^d} + q(u,v)_{L_2(\Omega)} \quad , \qquad u,v \in H^1(\Omega) .$$

We assume $0 < p_0 \leq p(\cdot) \leq p_1 < \infty$, and $q = const. > 0$ for simplicity.

Now we introduce a new unknown vector function: the flux $\mathbf{U} = -p \cdot \nabla u$ which usually has precise physical meaning depending on the application, e.g., it may repesent the current induced by the electrical potential u, or the particle flow in a porous medium driven by the pressure potential u etc.. The typical Hilbert space for $\mathbf{U} = (U_1, \ldots, U_d)$ is

$$H_0(\mathrm{div}; \Omega) \equiv \{\mathbf{U} \in (L_2(\Omega))^d \mid \mathrm{div}\,\mathbf{U} \in L_2(\Omega) \, , \, n \cdot \mathbf{U} = 0 \text{ on } \partial\Omega\} \, ,$$

with the norm given by

$$\|\mathbf{U}\|_{H(\mathrm{div})}^2 = \sum_{i=1}^{d} \|\mathbf{U}\|_{(L_2)^d}^2 + \|\mathrm{div}\,\mathbf{U}\|_{L_2}^2 \, .$$

We arrive at a new variational problem for $(\mathbf{U}, u) \in H_0(\mathrm{div}; \Omega) \times L_2(\Omega)$ (flux and potential):

$$\sum_i (1/p U_i, V_i)_{L_2} - (\mathrm{div}\,\mathbf{V}, u)_{L_2} = 0 \qquad \forall\, \mathbf{V} \in H_0(\mathrm{div}; \Omega)$$

$$(\mathrm{div}\,\mathbf{U}, v)_{L_2} + q(u, v)_{L_2} = (f, v)_{L_2} \qquad \forall\, v \in L_2(\Omega)$$

which can be discretized on its own (take an appropriate finite-dimensional subspace $\mathcal{V} \times \mathcal{W}$ of $H_0(\mathrm{div}; \Omega) \times L_2(\Omega)$) leading to a block-structured matrix problem

$$\begin{pmatrix} A & -B \\ B^T & qC \end{pmatrix} \cdot \begin{pmatrix} X \\ x \end{pmatrix} = \begin{pmatrix} 0 \\ b \end{pmatrix} \qquad .$$

with positive definite mass matrices A resp. C on the diagonal corresponding to the (weighted) L_2 scalar products on $\mathcal{V}$ resp. $\mathcal{W}$. As a whole, this is a non-symmetric saddle point problem. A typical strategy is to transform it to a symmetric problem by eliminating the $\mathcal{W}$ component. On the level of the continuous problem, this may be achieved by taking test-functions $v = 1/q \, \mathrm{div}\,\mathbf{V} \in \mathcal{W}$ where $\mathbf{V} \in \mathcal{V}$ in the second equation, and to add it to the first one. This results in

$$\sum_i (1/p U_i, V_i)_{L_2} + \frac{1}{q}(\mathrm{div}\,\mathbf{U}, \mathrm{div}\,\mathbf{V})_{L_2} = \frac{1}{q}(f, \mathrm{div}\,\mathbf{V})_{L_2}$$

to be satisfied for all vector functions $\mathbf{V} \in H_0(\mathrm{div}; \Omega)$. This is now a symmetric $H_0(\mathrm{div}; \Omega)$ elliptic problem for the flux variables. Thus, by the mixed method (which is actually a synonym for the transformed problems involving the new variable U and, in particular, for the above saddle point problem containing the old and new variables) we get direct approximations for both u and $\mathbf{U} = -p\nabla u$ which might be of physical interest. Computationally, the new formulations may be advantegeous since in order to get conforming subspaces $\mathcal{V}$ and $\mathcal{W}$ one has to satisfy less restrictive matching conditions on the inter-element boundaries which may positively influence the sparsity structure of the resulting discrete matrix problems. Anyhow, mixed formulations gain increasing popularity in a number of problems including Navier-Stokes equations and other flow problems [GiR, BF], plate and shell theory [BF, WT], Maxwell equations [KN, Mo1, Mo2] etc..

From the point of view of our approach, especially the symmetrized version (for the flux) of the mixed method is of interest. One may ask for appropriate (multilevel or other) splittings of $H_0(\mathrm{div};\Omega)$ or $H(\mathrm{div};\Omega)$. The same question can be posed for the spaces $H(\mathrm{curl};\Omega)$ serving the Maxwell-like systems. Since by now there is a great variety of general finite element constructions (RT(=Raviart-Thomas) elements, BDFM(= Brezzi-Douglas-Fortin-Marini) elements, Nedelec elements, and so on) for both triangular and rectangular partitions of practical interest in $\mathbf{R}^2$ and $\mathbf{R}^3$ applications, this might be an intriguing task, the more because the field is only beginning to attract the interest of the researchers. Some particular results are already available (cf. [EW1, VW1, CGP, Hn, Cw1, Cw2, Ku]) but we feel that much has still to be done in this direction. The typical idea for the $H(\mathrm{div};\Omega)$ situation is to separate first the divergence-free part

$$N^0(\mathrm{div};\Omega) = \{\mathbf{U} \in H(\mathrm{div};\Omega) \,|\, \mathrm{div}\,\mathbf{U} = 0\}$$

and to find a proper complement (not necessarily orthogonal) of $N^0(\mathrm{div};\Omega)$ in $H(\mathrm{div};\Omega)$. On the discrete level, for certain families of mixed finite elements, multilevel splittings of the corresponding two main parts of the subspace $\mathcal{V} \subset H_0(\mathrm{div};\Omega)$ have been proposed and discussed in [BP1, EW1] and in a series of subsequent papers by Ewing, Wang, Vassilevski et.al.. From the viewpoint of domain decomposition methods, splittings for the mixed finite element case have been considered in [Ma1, Ma2, Bn5]. We will not go into detail since the question needs more elaboration. For more background on various variants and applications of the mixed method and the above introductory remarks, we refer to the monographs and textbooks [BF, GiR, KN, Br1, Ci].

5 Error estimates and adaptivity

In this section, we come to further applications of approximation and function space theory to numerical multilevel schemes. Traditionally, we need them to justify the discretization process and to obtain qualitatively and quantitatively correct convergence estimates. In addition to the usual *a priori* error estimates (see briefly in 5.1), which depend on different regularity assumptions on the exact solution of the variational problem under consideration, one is equally interested in computable *a posteriori* estimates. The latter are important for adaptive processes (feedback control, as an example one can think about dynamic grid generation or refinement design). If such estimators are reliable and reflect the actual error (not the theoretical asymptotic upper bound which is typical for a priori error estimates) they may be of real importance for engineering applications. There is a huge volume of recent research literature about this topic, and we restrict our attention to a very particular but promising direction which was the starting point for us when getting involved in multilevel finite element approximation [Os1]. Such bounds can be treated within the framework of the approximation spaces $A_{p,q}^s(\{V_j\})$ introduced in 3.4. In 5.2 we prove some theoretical results on subspace selection related to the h-version of the finite element method (local grid refinement). The possible algorithmical consequences are discussed in 5.3. Another possibility to design an adaptive method based on the stable BPX-splittings of 4.2.1-2 is briefly mentioned in 5.3.

In the final subsection 5.4, we direct our attention to some unsolved nonlinear approximation problems related to those adaptive numerical methods in which the freedom of choice for the parameters of adaptivity is increased. Such methods choose adapted solutions from a complicated nonlinear manifold which contains qualitatively different atoms of approximation. The typical examples we have in mind include the h-p-version of the finite element method (regular local grid refinement plus arbitrary polynomial degree), the questions of wavelet packet approximation, and the problem of direction-dependent refinement naturally arising in the numerical modelling with nonlinear conservation laws and convection dominated diffusion. This subsection does not offer new results but rather informs the reader about an interesting research field.

5.1 Traditional error estimates

Finite element error estimates for variational problems are, in the most simple cases, consequences of the coercivity and boundedness of the bilinear form involved, and reduce to results on best approximation. E.g., if we look at the solution $u \in V$ of (**VP**) on an infinite-dimensional space V (see Section 4.1 where the precise definitions are given) and compare it to the approximate solutions $u_j \in V_j$ obtained by solving the same problem with respect to the finite-dimensional subspaces V_j, the Cea lemma reduces error estimates for $u - u_j$ in the energy norm $\| \cdot \|_E$ to estimates of best approximations

$$e_j(u)_E = \inf_{v_j \in V_j} \|u - v_j\|_E , \quad j = 0, 1, \dots , u \in V .$$

Thus, if we have a H^s-elliptic problem, and if we have more regularity for the exact solution u, we get correct bounds from results like Theorem 11 of section 3.4.2. The Aubin-Nitsche trick allows us to obtain results in some other norms (for second-order elliptic problems this is typically the L_2 norm, more generally H^s norms with $0 < s < 1$, which complements the energy resp. H^1 result). A third case of permanent interest is the derivation of estimates in the maximum norm $\|\cdot\|_{L_\infty}$. We refer to [Ci, Br1, Ha1, GR, BF, GiR], in these textbooks one can also find additional material on error analysis including the effects of boundary approximation and numerical integration resp. suitable for nonconforming and mixed methods which will be left out in our brief discussion.

A unifying general method covering the whole scale of Besov-Sobolev norms seems to be still missing despite the fact that over the past twenty years many research papers have been written on estimates in W_p^s norms, BMO etc. We refer to [CT, DNW, FR, GN, RSc, SF, Sc] for some typical investigations and methods. It is tempting to attack the general problem within the framework of $A_{p,q}^s$ spaces. We want to state this as an open problem which we leave up to the reader.

5.2 h-version and nonlinear approximation

This subsection is devoted to the theoretical background of the h-version of the finite element method. For a given elliptic problem (with specific right-hand side and boundary values) and fixed finite element type, the h-version aims at defining a partition $\mathcal{T}_n^*$ of the computational domain Ω into a fixed number n of subregions (triangles etc.) such that the approximation error of the finite element solution from $V(\mathcal{T}_n^*)$ is minimized in some suitable norm. Roughly speaking, this is the problem of optimizing the finite element method with respect to the geometry of the underlying partitions (in a first approximation, the computational work to find the finite element solution depends on $\dim V(\mathcal{T}_n^*) \approx |\mathcal{T}_n^*| \leq n$ which means that we minimize the discretization error without substantially increasing dimension of the computational subspace). During the 70's, the model problem to study the h-version from a theoretical and practical point of view was the problem of reentrant corners and interfaces in 2D elliptic boundary value problems, in the eighties the interest moved to investigations on adaptive finite element methods based on local a posteriori estimators as the most practical variant to find some (quasi-)optimal $\mathcal{T}_n^*$, see [BGO]. Especially for evolution problems where qualitative changes in the solution may occur in the course of time, error control and remeshing strategies are of practical interest.

We want to describe a more recent approximation-theoretical framework which shows under which assumptions the h-version with *regular* (but not necessarily quasi-uniform) triangulations gives generally better approximation rates compared with a standard finite element scheme based on a fixed sequence of quasi-uniform triangulations. As a by-product, we will get some sufficient conditions under which the use of nested refinement resp. nested selection of basis functions (see 4.2.2) in a multilevel context will increase the approximation power compared to a multilevel scheme without local refinement. The theoretical statements of this subsection will be discussed from a more practical point of view in the next subsection. We follow [Os1], compare also [DP2, DJP, Os2].

Let us fix a finite element approximation scheme satisfying the assumptions of Theorem 15 (if there is any problem with the understanding of the following discussion, think about linear triangular Lagrange elements in a two- or three-dimensional polyhedral domain). To simplify the representation, we will restrict ourselves to the classes $A_p^s \equiv A_{p,p}^s(\{V(\mathcal{T}_j)\})$, the general case can be found in [Os1]. Using exclusively the stability assumption, one can prove the following technical result which we will comment on afterwards. Note that we allow also for parameters $0 < p < 1$ (what we could have done from the very beginning, see [Os1, DJP]).

Theorem 21. Let $f \in A_p^s$ for some $0 < p < \infty$ and $s > 0$. Suppose that the parameters p', s' satisfy $p < p' \leq \infty$ and $0 < s' < s - d(1/p - 1/p')$ (this, for instance, guarantees that A_p^s is compactly imbedded into $A_{p'}^{s'}$). Then for all sufficiently large n there is a finite linear combination of $\leq n$ basis functions

$$h_n^* = \sum_{j=0}^{\infty} \sum_i c_{j,i}^* N_{j,i} , \quad \operatorname{card}\{(j,i) \,:\, c_{j,i}^* \neq 0\} \leq n ,$$

such that

$$\|f - h_n^*\|_{A_{p'}^{s'}} \leq C n^{-(s-s')/d}\|f\|_{A_p^s}$$

and

$$\|h_n^*\|_{A_p^s} \leq C n^{-(s-s')/d}\|f\|_{A_p^s} .$$

Analogously, one can show the existence of a regular partition $\mathcal{T}_n^*$ into $\leq n$ elements and of a finite element function $h_n^* \in V(\mathcal{T}_n^*)$ satisfying the same two estimates.

Proof. Let us take an A_p^s-admissible representation

$$f = \sum_{j=0}^{\infty} g_j , \quad g_j = \sum_i c_{j,i} N_{j,i} \in V_j ,$$

satisfying

$$\|\{2^{sj}\|g_j\|_{L_p}\}\|_{l_p} \leq C\|f\|_{A_p^s} .$$

Fix a natural number J such that $n \approx 2^{Jd}$. We use the following *selection rules* depending on two parameters ϵ and $\tilde{c}$ which will be fixed later on:

$$c_{j,i}^* = \begin{cases} 0 & \text{if } j > J \text{ and } |c_{j,i}| < \tilde{c}2^{-j\epsilon} \\ c_{j,i} & \text{elsewhere} \end{cases}$$

Roughly speaking, besides the $\approx 2^{Jd}$ basis functions belonging to levels $j \leq J$ we select some more from higher levels $j > J$ according to the size of their coefficients.

Choosing the right ϵ and $\tilde{c}$, we will show that $h = \sum_j \sum_i c_{j,i}^* N_{j,i}$ has (almost) the desired properties. Since a representation of h results from that of f by dropping terms one obviously has

$$\|h\|_{A_p^s}^p \leq C \sum_j \sum_i 2^{j(sp-d)}|c_{j,i}^*|^p \leq C \sum_j \sum_i 2^{j(sp-d)}|c_{j,i}|^p \leq C\|f\|_{A_p^s}^p .$$

The following argument is carried out for $p' < \infty$, the changes for $p' = \infty$ are left to the reader. Looking at the formal representation of $f - h$ resulting from our selection rules, we get

$$\begin{aligned} \|f - h\|_{A_{p'}^{s'}}^{p'} &\leq C \sum_{j>J} \sum_{i:|c_{j,i}|<\tilde{c}2^{-j\epsilon}} 2^{j(s'p'-d)}|c_{j,i}^*|^{p'} \\ &\leq C \sum_{j>J} \sum_i 2^{j(s'p'-d)}(\tilde{c}2^{-j\epsilon})^{p'-p}|c_{j,i}|^p \end{aligned}$$

$$\leq \; C\tilde{c}^{p'-p} \sum_{j>J} 2^{j(s'p'-\epsilon(p'-p))} \|g_j\|_{L_p}^p$$

$$\leq \; C\tilde{c}^{p'-p} 2^{J(s'p'-sp-\epsilon(p'-p))} \Big(\sum_{j>J} 2^{jsp} \|g_j\|_{L_p}^p \Big)$$

whenever ϵ is taken such that $s'p' - sp - \epsilon(p'-p) \leq 0$. If we fix

$$\tilde{c} = 2^{J(\epsilon-s)} \Big(\sum_{j>J} 2^{jsp} \|g_j\|_{L_p}^p \Big)^{1/p}$$

we get the second inequality:

$$\|f-h\|_{A_{p'}^{s'}}^{p'} \leq C 2^{J(s'-s)p'} \Big(\sum_{j>J} 2^{jsp} \|g_j\|_{L_p}^p \Big)^{p'/p} \leq C n^{-(s-s')p'/d} \|f\|_{A_p^s}^{p'} \; .$$

It remains to check that the number of non-zero coefficients $c_{j,i}^*$ in the representation of h is of order n. Let n_j^* denote the number of non-zero coefficients for fixed $j = 0, 1, \ldots$. Obviously, $n_j^* \leq \dim V_j \leq C 2^{jd}$ for $j \leq J$, and

$$n_j^* \leq \sum_i (\tilde{c} 2^{-j\epsilon})^{-p} |c_{j,i}|^p \leq C \tilde{c}^{-p} 2^{j(d+\epsilon p)} \|g_j\|_{L_p}^p \;, \; j > J \; .$$

Summing up, we have

$$\mathrm{card}\,\{(j,i) \; : \; c_{j,i}^* \neq 0\} = \sum_{j=0}^{\infty} n_j^* \leq C \Big(2^{Jd} + \tilde{c}^{-p} 2^{J(d+(\epsilon-s)p)} \sum_{j>J} 2^{jsp} \|g_j\|_{L_p}^p \Big)$$

if we assume $d + (\epsilon - s)p \leq 0$ (as a matter of fact, the two conditions on ϵ can be satisfied simultaneously since the resulting condition takes the form

$$\epsilon \in \Big[s + \frac{(s'-s)p'}{p'-p}, s - \frac{d}{p} \Big] \,,$$

with a non-empty interval on the right due to the assumptions on s' and p'). Thus, by the definition of $\tilde{c}$ and of J we finally get

$$\mathrm{card}\,\{(j,i) \; : \; c_{j,i}^* \neq 0\} \leq C 2^{Jd} \leq C n \; .$$

Since the constants do not depend on n, it is a simple scaling argument which leads to a number of $\leq n$ non-zero coefficients (n sufficiently large) in the construction and to the desired h_n^*. The two inequalities still hold for h replaced by the latter, with possibly larger constants.

Finally, the assertion on the existence of suitable partitions into $\leq n$ elements is a consequence of the result just proved (for triangulations this has been checked in detail in [Os1], the argument leads even to a partition which may be obtained by nested refinement as discussed in 4.2.2). Theorem 21 is proved.

A few comments on Theorem 21 are in order. First of all, using Theorem 6 and the usual tricks (cf. the remarks following Theorem 11) one. may reformulate the above result in the language of Besov-Sobolev spaces. We leave this as an exercise up to the reader, some special cases are mentioned below.

If we introduce the set

$$V_n^* = \{h_n^* = \sum_j \sum_i c_{j,i}^* N_{j,i} \ : \ \mathrm{card}\{(j,i) \ : \ c_{j,i}^* \neq 0\} \leq n\} \subset L_\infty(\Omega)$$

and denote the best approximations of a function $f \in Y$ with respect to this nonlinear set and some norm $\| \cdot \|_X$ by $e_n^*(f)_X$ we actually have proved the following Jackson type estimate

$$e_n^*(f)_{A_{p'}^{s'}} \leq C n^{-(s-s')/d} \|f\|_{A_p^s} \qquad \forall f \in A_p^s$$

(with the range of parameters as given above). Alternatively, we could have written down an upper estimate also for the best approximations $\tilde{e}_n(f)_{A_{p'}^{s'}}$ with respect to the set $\tilde{V}_{n,reg}$ of all finite element functions corresponding to regular partitions into $\leq n$ elements. In contrast to what we have discussed in Section 2, these are typical results of nonlinear approximation theory since V_n^* is nonlinear (as a matter of fact, we have $V_n^* + V_n^* \subset V_{2n}^*$, and the factor 2 is best possible). We refer, e.g., to [Br2, PP, De2, LD] for some introduction into this field.

It is fair to compare the above estimate with estimates for $E_{V_j}(f)_{A_{p'}^{s'}}$ ($n \approx \dim V_j \approx 2^{dj}$) resp. for $E_{V(T)}(f)_{A_{p'}^{s'}}$ ($|T| \approx n$). Obviously,

$$e_{\dim V_j}^*(f)_{A_{p'}^{s'}} \leq E_{V_j}(f)_{A_{p'}^{s'}}$$

for any reasonable f since $V_j \subset V_{\dim V_j}^*$, and the question is whether the enlargement of the set of approximating functions gives a substantial improvement. By Theorem 11 and a simple imbedding argument one has

$$E_{V_j}(f)_{A_{p'}^{s'}} \leq C n^{-(s-s')/d+(1/p-1/p')} \|f\|_{A_p^s} \qquad \forall f \in A_p^s \quad (n \approx \dim V_j)$$

for the same range of parameters. We see that we have lost a factor $n^{-(1/p-1/p')}$ which shows that nonlinear approximation may lead to results better by an order. We will further illustrate this below on the reentrant corner problem for polyhedral domains.

Note once again that the theory of $A_{p,q}^s$ spaces may be generalized to values $p,q < 1$ as well, and that this extension is important in the present discussion, see [Os1] or [De2]. E.g., we have the following corollary (for $p < 1$, L_p and the quasi-normed Besov spaces $B_p^{s,m} = B_{p,p}^{s,m}(\Omega)$ are defined in the same way as for $p \geq 1$):

$$e_n^*(f)_{L_{p'}} \leq Cn^{-s/d}\|f\|_{B_p^{s,m}} \qquad \forall f \in B_p^{s,m}$$

holds for the range of parameters $0 < p < p' \leq \infty$, $d(1/p - 1/p') < s < m$, with m given in Theorem 6.

Moreover, as is shown by a very deep and nice result of [DJP], the role of Besov spaces with smaller $p < p'$ (often $p < 1$) is crucial in this type of nonlinear approximation (for some history of this result, see [De2]): E.g., for the corresponding approximation scheme with smooth tensor product splines of order m (the case $m = 2$ corresponds then to bilinear finite elements) in a d-dimensional cube, one has

$$\sum_{n=1}^{\infty} (n^{s/d}e_n^*(f)_{L_{p'}})^p \frac{1}{n} < \infty \quad \Longleftrightarrow \quad f \in B_p^{s,m}$$

if $s = d(1/p - 1/p')$ belongs to the open interval $(0, m)$. This is a rare case in nonlinear approximation: the assertion describes the exact class of functions with a given (or desired) rate of nonlinear approximation. It also shows that Theorem 21 approaches the optimum one can get (actually, Theorem 21 gives sufficient conditions for a wider range of parameters, and the limiting case $s - s' = d(1/p - 1/p')$ excluded there corresponds to the necessary and sufficient condition just stated in a special case). To prove such results one needs the above Jackson type estimate (with $A_{p'}^{s'}$ replaced by $L_{p'}$) and a more involved Bernstein type inequality. A slightly different method has been proposed in [JaM] for the case of nonlinear wavelet approximation in Besov spaces.

Next, one might ask about the efficient construction of a (quasi-)best h_n^* resp. $\mathcal{T}$ with $|\mathcal{T}| \leq n$. The proof of Theorem 21 suggests rules one could follow if a suitable, optimal decomposition of f in A_p^s would be available (roughly speaking, the rule would be to select the terms with large coefficients by choosing a proper bound for what means "large", cf. the theoretical rules used in the proof). The weak point, however, is that often the decomposition is not available or too expensive to obtain. Algorithms have been successfully tested for surface resp. image compression (pixel pictures) [DJ1, DJ2] where discrete wavelet decompositions may be obtained at relatively low costs. On the contrary, for p.d.e. solving it is not realistic to assume that some decomposition of the unknown solution is at hand. Here, the only way is the combination with adaptive schemes including additional error estimators. We come back to this point in the next subsections. In the theory of approximation, such more realistic procedures are classified under the name adaptive approximation, see [De2, DY, Sr2] for some results and references.

We finish with a particular application of Theorem 21 resp. the results of [Os1] to the corner singularity problem for 2D elliptic problems (for details, see [Os1], section 5). Let us consider the Poisson equation with homogeneous mixed Dirichlet-Neumann boundary conditions

$$-\Delta u = f\,, \qquad u|_{\Gamma_D} = 0\,, \qquad \frac{\partial u}{\partial n}\Big|_{\Gamma_N} = 0\,,$$

in a plane polygonal domain Ω with boundary $\partial\Omega = \Gamma_D \cup \Gamma_N$ ($\Gamma_D \cap \Gamma_N = \emptyset$, $\Gamma_D \neq \emptyset$). We suppose that $f \in L_p$ for some $1 < p \leq 2$. Then, by regularity results quoted in [Gv, Dg], the solution admits a splitting $u = u_{reg} + u_{sing}$ into a regular part $u_{reg} \in W_p^2(\Omega)$ and a singular part consisting of a sum of special singularity functions S_j each corresponding to some corner point:

$$u_{sing} = \sum a_j S_j\,, \qquad S_j(x) = \phi_j(x) r^{\lambda_j}(\log r)^{m_j}$$

with r denoting the distance of x from the corresponding corner point, $m_j = 0$ or $m_j = 1$, $1/4 \leq \lambda_j \leq 2 - 2/p$ depending on the interior angle, and $\phi_j(x)$ a smooth cut-off function with support in a neighbourhood of this corner which satisfies the homogeneous boundary conditions. See Figure 20 for an illustration. Large

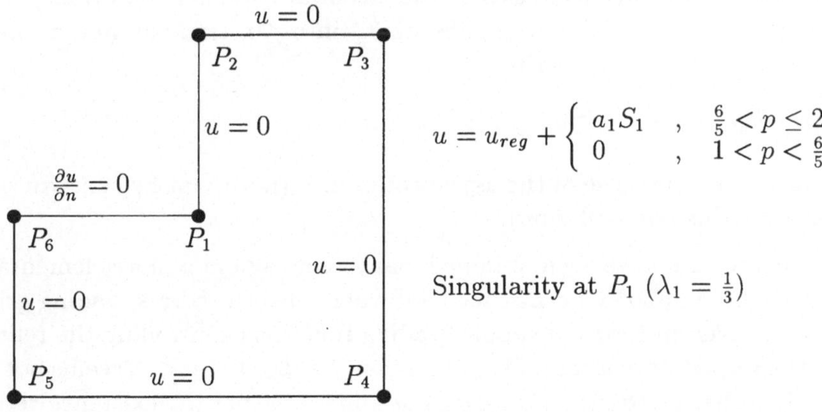

$$u = u_{reg} + \begin{cases} a_1 S_1 & , \quad \frac{6}{5} < p \leq 2 \\ 0 & , \quad 1 < p < \frac{6}{5} \end{cases}$$

Singularity at P_1 ($\lambda_1 = \frac{1}{3}$)

Figure 20. Mixed boundary value problem in a L-shaped domain

interior angles may therefore cause singularities which lower the overall smoothness (i.e. $u \in W_p^s$ only for $s < s_0 = \min(\lambda_j + 2/p)$) and contribute to the well-known pollution effect (finite element approximations on quasi-uniform triangulations become of lower order accuracy even far away from such dangerous corner points).

Anyhow, u is still in $H^1_{D=\Gamma_D}(\Omega)$ and one may ask about improvements if one adapts the triangulation to resolve the singularities (another standard idea to avoid the pollution effect is the use of special elements modelling the form of the singularity functions S_j which is delicate if the boundary value problem is more general than the above one or if the domain is three-dimensional). Theorem 21 (more exactly, its generalization to $A^s_{p,q}$-spaces discussed in [Os1]) contains an existence result which answers this question. Consider linear triangular finite elements and suppose that the initial triangulation is compatible with the two parts Γ_D and Γ_N of the boundary. This allows us to define the $A^s_{p,q,D=\Gamma_D}$-spaces with respect to the sequence $V_{j,D=\Gamma_D}$ of linear f.e. functions on $\mathcal{T}_j$ vanishing on Γ_D, see 3.4.4. Checking the exact smoothness in terms of Besov norms of the singularity functions S_j one gets $S_j \in B^{2,2}_{\tilde{p},\infty;D=\Gamma_D}$ for $\tilde{p} < 2/(2 - \lambda_j)$ and taking into account the restrictions on λ_j and p, we arrive at

$$u \in B^{2,2}_{\tilde{p},\infty;D=\Gamma_D} \subset A^2_{p,\infty}(\{V_{j,D=\Gamma_D}\})\,, \qquad 1 < \tilde{p} < \min(p, 8/7)\,.$$

Since this $A^s_{p,q}$-space is (compactly) imbedded into $H^1_{D=\Gamma_D}(\Omega) = A^1_{2,2}(\{V_{j,D=\Gamma_D}\}) = A^1_{2,2;D=\Gamma_D}(\{V_j\})$, we may apply a variant of Theorem 21 (we actually need the case $p = \tilde{p}$, $q = \infty$, $s = 2$ on the one hand, $p' = q' = 2$, $s' = 1$, on the other, and the obvious modifications to satisfy the essential Dirichlet conditions) that gives the existence of a regular triangulation $\mathcal{T}_n^*$ into $\leq n$ triangles and a linear f.e. function $h_n^* \in V(\mathcal{T}_n^*)_{D=\Gamma_D}$ with

$$\|u - h_n^*\|_{H^1} \leq C n^{-1/2} \|u\|_{B^{2,2}_{\tilde{p},\infty}}\,, \qquad n \to \infty$$

which is optimal in the sense of the asymptotics in n (note that by construction $h(\mathcal{T}_n^*) \approx n^{-2}$ for this triangulation).

For $p = 2$ such results have been obtained many years ago in a more elementary way using weighted Sobolev norms (see [Gv] where also the corresponding grids are explicitely given in terms of simple grading functions controlling the refinement into the respective corner point). The case $f \in L_p$, $1 < p < 2$, seems not to be covered by other methods, and we do not know whether the existence result may be turned into a practical method achieving the asymptotical optimal rate in this case.

5.3 Adaptive multilevel methods

We continue the discussion of the h-version of the finite element method, and turn to the more practical question of how to organize it on a computer. The results of the previous subsection show what kind of theoretical a priori assumption would lead to improved approximation rates, and that such rates may be achieved by nested subspace selection resp. nested refinement. On the other hand, from section 4.2.2 we already know that linear systems resulting from a multilevel finite element discretization including such type of refinement may be solved with almost the same efficiency as in the case of regular dyadic refinement. What remains is to find implementable criteria for controlling the refinement process using a minimum of additional information. There is no general answer to this question at present but a growing body of proposals and mathematical and engineering literature, see [BGO, TWM, Ve2]. Since we are slaves of the decomposition approach we want to emphasize the way we would like to approach the problem.

A first simple variant is as follows. As was pointed out before, the main problem in finding the optimal h_n^* of Theorem 21 is that the *whole* decomposition of f is not available. Instead, we might try to find only a reasonable part of it. To make the idea precise, suppose that we have a suitable multilevel scheme

$$V_0 \subset V_1 \subset \ldots \subset V_j \subset \ldots$$

of f.e. subspaces on regularly refined partitions which allows for our $A_{p,q}^s$-theory. Suppose that the solution operators $P_j : V \to V_j$ of the discrete variational problems (with V typically coinciding with some $A_{2,2}^s(\{V_j\})$) give rise to stable decompositions

$$u = \sum_{j=0}^{\infty} \tilde{g}_j \equiv \sum_{j=0}^{\infty} (P_j u - P_{j-1} u)$$

in the sense that

$$\|\{2^{js}(P_j u - P_{j-1} u)\}\|_{l_q(L_p)} \approx \|u\|_{A_{p,q}^s} \qquad \forall\, u \in A_{p,q}^s(\{V_j\})$$

for a certain nontrivial range of parameters p, q, s $(P_{-1} \equiv 0)$.

Suppose that we have found $u_j \equiv P_j u$ for $j \le j_0$. Now we want to judge the quality of the last approximation u_{j_0} resp. to decide whether to fully or partly compute more u_j (with $j > j_0$). According to our theory, the complete answer would follow from the knowledge of the coefficients $c_{j,i}$ in

$$\sum_{j=j_0+1}^{\infty} (u_j - u_{j-1}) = \sum_{j=j_0+1}^{\infty} c_{j,i} N_{j,i}$$

which we are not able to compute. What we propose to try is to *prove* that under certain assumptions (which should be satisfied for at least some problem classes of practical importance) it is sufficient to find a few new approximations $u_{j_0+1}, \ldots, u_{j_0+k_0}$, to obtain from them an incomplete decomposition

$$\sum_{j=j_0+1}^{j_0+k_0} (u_j - u_{j-1}) = \sum_{j=j_0+1}^{j_0+k_0} \sum_i c_{j,i} N_{j,i} \, ,$$

and to use it to select the new, adapted subspace $V_{j_0+1}^* \subset V_{j_0+1}$ by the method of large coefficients as described in 5.2. The k_0 should be kept small but not too small in order to have a still cheap and robust selection procedure.

Clearly, after having fixed this idea of selecting a next subspace, it remains for us to work out many details (especially, one might be interested in restricting the construction in such a way that the resulting sequence of optimally chosen subspaces

$$V_0^* \subset V_1^* \subset \ldots \subset V_j^* \subset \ldots$$

fits the discussion of 4.2.2 in order to have uniform estimates to control the selection process for all j). A good test case would be once again the (already settled) question of corner singularities where we see the chance to get both theoretical results and practical comparisons with other well-established methods. Even if one can not expect a full theoretical justification in more complicated situations, one has enough background to develop and test a code implementing the above strategies.

Let us point out that the above algorithmical idea is not new, it has been proposed many times before. We refer, e.g., to work reported on in [Ve1, Ve2]. The difference is that due to 4.2.2 and 5.2 we have a specific background to justify and further develop this idea. A one-dimensional variant of this approach has been discussed in [Os13], some later sporadic numerical experiments have supported the basic idea. There are, however, some simple arguments against the above procedure: one may look at it as a specific way to construct a local a posteriori error estimator which is obtained by taking differences of the actual approximation with the following ones of the *same* type and, thus, of the same order of approximation (our schemes are linearly convergent). This may cause the danger of underestimating the error, and lower the practical reliability of the selection process. In contrast, the well-established a posteriori error estimation techniques of Babuska-Rheinboldt-Miller and Zienkiewicz-Zhu but also those used in PLTMG or Kaskade explore higher-order approximation (i.e. an auxiliary quadratic f.e. approximation if linear elements are used for the algorithm itself), compare [BGO, ZZ, BSW, Ba, Ln, Ro, BE2, BE1]. We come back to this point a bit later.

We want to state another simple consequence of the special decomposition norms implicitly contained in Theorem 19. For simplicity, let us consider the MDS splitting for linear triangular f.e., applied to a variational problem given by a symmetric H^1-elliptic bilinear form $a(\cdot,\cdot)$. Then, Theorem 19 states

$$a(u,u) \approx a(P_{\sum V_j} u, u) = \sum_{j=0}^{\infty} \sum_i \frac{a(u, N_{j,i})^2}{a(N_{j,i}, N_{j,i})} \qquad \forall\, u \in V = H^1(\Omega) \, .$$

Taking now an arbitrary conforming approximation $u^* \in V$ to u, we get an estimate of the energy norm of the error $e = u - u^*$:

$$\|e\|_E^2 \approx \sum_{j=0}^{\infty} \sum_i \frac{r_{j,i}^2}{d_{j,i}}$$

where $d_{j,i} = a(N_{j,i}, N_{j,i})$, and

$$r_{j,i} = a(e, N_{j,i}) = \Phi(N_{j,i}) - a(u^*, N_{j,i})$$

is the residual of the approximate solution u^* when testing the variational problem with the basis function $N_{j,i}$. If $u^* = P_j u$ is the (exact) discrete solution on V_j then the residuals $r_{l,i}$ are zero for all $l \leq j$, and the remaining error is the sum of the squares of scaled residuals of test equations corresponding to basis functions $N_{l,i}$ with $l > j$ having small local support. This immediately leads to theoretical a posteriori estimates

$$\|u - P_j u\|_{H^1(\Omega)}^2 \leq C \sum_{l>j} \sum_{i\,:\,\mathrm{supp}\,N_{l,i}\cap G\neq\emptyset} \frac{r_{l,i}^2}{d_{l,i}}$$

which can be replaced in a more or less justified way by computable ones (e.g. using certain decay properties of the $r_{l,i}$ obtained from additional regularity assumptions on u).

Note that these considerations are possible for all kind of stable decompositions of the *infinite-dimensional* solution space V. E.g., the BPX splitting or pre-wavelet schemes are good candidates, too. Some work on adaptivity within the wavelet decomposition framework and theorems like Theorem 21 has recently been carried out by DeVore, Lucier, and Jawerth [DJ1, DJ2], with typical applications to image and surface compression. Another interesting possibility is as follows. Denoting, for a given sequence of partitions of a polyhedral domain, the subspaces of linear and quadratic elements by $\{V_j\}$ and $\{\hat{V}_j\}$, resp., we may consider a whole family of subspace splittings of $V = H^1(\Omega)$ given by

$$V_0 \subset \ldots \subset V_j \subset \hat{V}_j \subset \hat{V}_{j+1} \subset \ldots$$

According to the discussion in 4.2.3, these splittings possess condition numbers of the corresponding additive Schwarz operators that are uniformly bounded in j. The same is true for the refined splittings (of BPX or MDS type) which will be obtained if the usual nodal basis $\{N_{j,i}\}$ is taken for V_j, and a basis of $\hat{V}_j$ is obtained by complementing that of V_j by piecewise quadratic bubble functions $\hat{N}_{j,e}$ corresponding to the midpoints of the edges e of the partition $\mathcal{T}_j$ (this results in the so-called hierarchical (in the spirit of Zienkiewicz [ZM]) basis for $\hat{V}_j$). A moment's reflection shows that

$$\|u_j - \hat{u}_j\|_E^2 \approx \sum_{\text{edges of } \mathcal{T}_j} \frac{(\Phi(\hat{N}_{j,e}) - a(u_j, \hat{N}_{j,e}))^2}{a(\hat{N}_{j,e}, \hat{N}_{j,e})} \, , \quad u_j = P_j u \, , \ \hat{u}_j = \hat{P}_j u$$

uniformly in j where $\hat{P}_j : V \to \hat{V}_j$ denotes the corresponding elliptic projection onto the subspaces of quadratic elements. Now, if the exact solution u is sufficiently smooth we may assume that $\|u - u_j\|_E \approx \|u_j - \hat{u}_j\|_E$, with constants in the corresponding two-sided inequalities that are close to 1. This reflects the assumption that the quadratic approximation is by an order better than the linear one. Therefore, the above expression yields a computable error estimator which is similar to the edge-oriented error estimators used in Kaskade [DLY, BE1]. More important than this particular coincidence is the general possibility to design various local a posteriori estimators from the stock of stable multilevel splittings for Sobolev and other Hilbert spaces used in p.d.e. theory. We feel that the relationship of optimal multilevel splittings on the one hand, and adaptivity concepts and computable error estimators, on the other, deserves further attention and research.

We finish with briefly mentioning one more recent activity in this direction, the so-called fully adaptive multilevel method by Rüde [Rd1, Rd3]. A description of the basic idea, to consider iteration, error estimation, and adaptive refinement as integrated parts of a Gauss-Seidel type iteration procedure on the partially activated infinite-dimensional $V = \sum_j \sum_i V_{j,i}$, along the lines of our approach is given in [Os10]. We refer to analogous developments in signal and wavelet processing [MZ].

5.4 More complicated approximation schemes

What we currently meet in some adaptive numerical methods is the increased complexity of the approximating nonlinear manifold. We want to describe some examples of practical interest that show the problems, and will propose some strategy to develop a theory which is close to the needs of a real-life adaptive process but hopefully allows for some theory in the sense of 5.2 (i.e. for sufficient conditions on a problem under which adaptivity gives substantially better rates of approximation and therefore makes sense to be applied). In comparison with the theory of subsection 5.2 where the starting point was a *linearly* increasing sequence of subspaces

$$V_0 \subset V_1 \subset \ldots \subset V_j \subset \ldots ,$$

each equipped with well-localized, L_p-stable algebraic bases, and possessing a theory of $A_{p,q}^s$-spaces, we want to work now in a more complicated framework of an at least two-parameter family of subspaces V_{j_1,j_2} or, more generally, in a tree-structured family of subspaces given by a father-son mapping $V_\alpha \to \Sigma(V_\alpha) = \{V_\beta : \beta \in \sigma(\alpha)\}$ where σ is a mapping on the set of indices which represents the tree and corresponds to the data structure behind. Figures 21 and 22 schematically show these two models. Note that the scheme shown in Figure 21 is a particular case of the second scheme (consider the mapping $\sigma((j_1,j_2)) = \{(j_1+1,j_2),(j_1,j_2+1)\}$). We generally assume that

$$V_\alpha \subset V_\beta \qquad \forall\,\beta \in \sigma(\alpha)$$

and that some sort of nodal basis may be chosen (here we have to allow for more flexibility than before to cover some realistic applications).

The *more complicated* adaptive method we have in mind consists in, first, the selection of a *path*

$$V_{\alpha_0} \subset V_{\alpha_1} \subset \ldots \subset V_{\alpha_j} \subset \ldots \quad , \qquad \alpha_{j+1} \in \sigma(\alpha_j) \quad \forall\,j \geq 0.$$

and after this in a second selection of subspaces $V_j^* \subset V_{\alpha_j}$ within the path, say, following the strategies of 5.2. The first step may be called *global selection* while the second corresponds to *local selection* or *local refinement*.

Algorithmically, there should not be a strict difference between the two stages but we hope that we get some benefit in theory: if we would be able to develop a $A_{p,q}^s$-theory (or other theory of suitable decomposition-like spaces) for each path separately, and to describe the resulting function spaces in terms allowing a treatment by the usual elliptic regularity theory, we will eventually have some

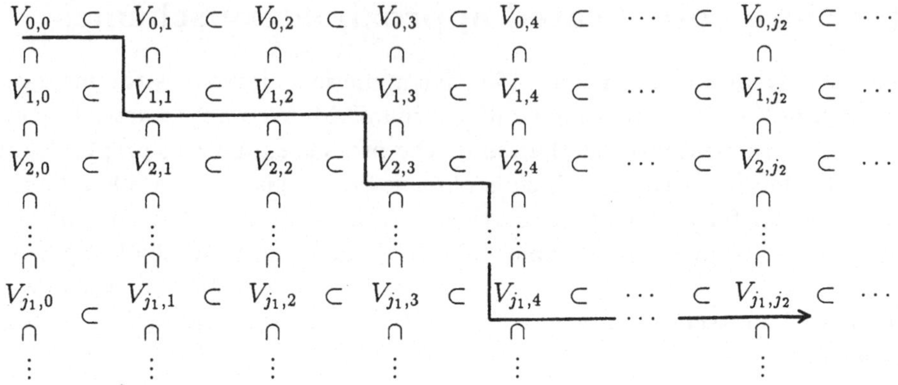

$$
\begin{array}{ccccccccccccc}
V_{0,0} & \subset & V_{0,1} & \subset & V_{0,2} & \subset & V_{0,3} & \subset & V_{0,4} & \subset & \cdots & \subset & V_{0,j_2} & \subset & \cdots \\
\cap & & \cap & & \cap & & \cap & & \cap & & & & \cap \\
V_{1,0} & \subset & V_{1,1} & \subset & V_{1,2} & \subset & V_{1,3} & \subset & V_{1,4} & \subset & \cdots & \subset & V_{1,j_2} & \subset & \cdots \\
\cap & & \cap & & \cap & & \cap & & \cap & & & & \cap \\
V_{2,0} & \subset & V_{2,1} & \subset & V_{2,2} & \subset & V_{2,3} & \subset & V_{2,4} & \subset & \cdots & \subset & V_{2,j_2} & \subset & \cdots \\
\cap & & \cap & & \cap & & \cap & & \cap & & & & \cap \\
\vdots & & \vdots & & \vdots & & \vdots & & \vdots & & & & \vdots \\
\cap & & \cap & & \cap & & \cap & & \cap & & & & \cap \\
V_{j_1,0} & \subset & V_{j_1,1} & \subset & V_{j_1,2} & \subset & V_{j_1,3} & \subset & V_{j_1,4} & \subset & \cdots & \subset & V_{j_1,j_2} & \subset & \cdots \\
\cap & & \cap & & \cap & & \cap & & \cap & & & & \cap \\
\vdots & & \vdots & & \vdots & & \vdots & & \vdots & & & & \vdots
\end{array}
$$

Figure 21. Two-parameter subspace scheme

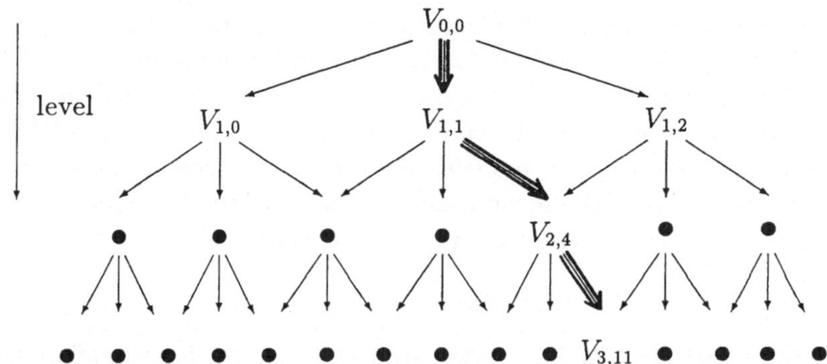

Figure 22. Path in a subspace tree scheme

qualitative and quantitative understanding of the approximation power and the range of applicability of the resulting adaptive method. The implementation, clearly, requires the same mathematically doubtful tricks as before. The following examples do not give any definitive results but rather illustrate our point of view, and should play the role of an appetizer to interest the reader in nonlinear approximation problems and more complicated adaptive schemes.

5.4.1 The h-p-version

In potential theory and for elasticity problems, one often has the situation that the solution is piecewise analytic, except some singular behavior near corner points, edges, cuts resp. cracks or some other lower-dimensional boundary parts. In addition, interfaces may occur due to different material constants if composites are considered. Knowing about the exclusive role of polynomials for approximating analytic functions, and about the potential of the h-version to resolve singularity parts by local refinement, it is a natural step to combine the two approaches. The result is the h-p-version which may be characterized roughly as follows: in regions of analyticity high order polynomials are used over relatively large subdomains, while near the singularity manifolds the preference is given to strongly refined partitions and low order polynomials. It turns out that for certain classes of functions this combination of the pure h-version (keeping the local polynomial degree fixed but the partition variable) and the pure p-version (piecewise polynomial approximation of arbitrary degree on a fixed partition) is superior over each of these two basic methods.

To keep the exposition elementary, let us consider the one-dimensional situation, and a conforming approach to second order elliptic problems (cf. [GB]). Let $(\pi, \mathbf{r}) \in \mathcal{I}_n$ if the partition π of $[0,1]$ is given by $m \geq 0$ interior knots $(x_0 =)0 < x_1 < \ldots < x_m < 1(= x_{m+1})$, and the integer vector $\mathbf{r} = (r_0, \ldots, r_m)$ satisfies $r_i \geq 1$ and $\sum_{i=0}^{m} r_i < n$. Consider the subspace

$$S_{\mathbf{r}}^0(\pi) = \{g \in C[0,1] \mid g|_{[x_i, x_{i+1}]} \text{ is a polynomial of degree} \leq r_i\}$$

of univariate C^0 f.e. functions of variable degree. By construction, dim $S_{\mathbf{r}}^0(\pi) \leq n$ for any $(\pi, \mathbf{r}) \in \mathcal{I}_n$. We define three best approximations for $f \in X = H^1(0,1)$ which correspond to the h-version of the finite element method with type r Lagrange elements, to the p-version over an arbitrarily fixed partition π, and to the h-p-version, resp.:

$$e_n^{(r)}(f)_X = \inf_{\pi\,:\,(\pi,(r,\ldots,r))\in\mathcal{I}_n} \inf_{g\in S_{(r,\ldots,r)}^0(\pi)} \|f - g\|_X$$

$$e_n^{(\pi)}(f)_X = \inf_{\mathbf{r}\,:\,(\pi,\mathbf{r})\in\mathcal{I}_n} \inf_{g\in S_{\mathbf{r}}^0(\pi)} \|f - g\|_X$$

$$e_n(f)_X = \inf_{(\pi,\mathbf{r})\in\mathcal{I}_n} \inf_{g\in S_{\mathbf{r}}^0(\pi)} \|f - g\|_X .$$

Clearly, one may study these nonlinear approximation numbers also for approximation in other norms, e.g., in $X = L_p(0,1)$. In a remarkable case study, Gui and Babuska [GB] examined these quantities for $f_\alpha(x) = x^\alpha$, $x \in [0,1]$, where $\alpha > 1/2$ to ensure that $f_\alpha \in H^1(0,1)$. This type of function models (in dimension one) the typical singularity behavior near corner points for 2D problems with analytic data. A typical result of [GB] is summarized by the following sharp asymptotic estimates:

$$e_n^{(r)}(f_\alpha)_X = O(n^{-r}) \,,$$

$$e_n^{(\pi)}(f_\alpha)_X = O(n^{-2(\alpha-1/2)}) \,,$$

$$e_n(f_\alpha)_X = O(e^{-(c\sqrt{(\alpha-1/2)n})}) \,, \ c \approx 1.76 \,,$$

as $n \to \infty$. For completeness, let us note that the asymptotic rates for the finite element approximation of f_α by Lagrange elements of degree r on a sequence of uniform partitions are of order $O(n^{-\min(\alpha-1/2,r)})$, with n once again indicating the dimension of the corresponding approximating subspace.

These rates show the completely different nature of the h-p-approximation. Analogous results have been obtained by Gu, Babuska [BG] for 2D problems (these authors also provided new elliptic regularity results for classes of piecewise analytic data which indicate the type of spaces one might look for if characterization theorems for nonlinear h-p-approximation are required). At present, work on 3D applications, on applications of the same idea within the concepts of the boundary element method etc. is going on, see [Bb1, Bb2, BS, BGS, El]. On the other hand, the approximation-theoretical background of the method is still not sufficiently clear.

We come back to the scheme we have introduced at the beginning of section 5.4. First of all, in analogy to 5.2 we restrict the flexibility of grid selection by fixing the sequence of dyadic uniform partitions $\{\pi_k^*\}$ as the basic one (more precisely, π_k^* is uniform with stepsize 2^{-k}, $k = 0, 1, \ldots$). We define $V_{r,k}$ as the subspaces of univariate C^0 Lagrange elements of degree r (order $m = r+1$) on π_k^*, i.e. $V_{r,k} = S^0_{(r,\ldots,r)}(\pi_k^*)$, $k \geq 0$, $r \geq 1$. Thus, we arrive at a two-parametric approximation scheme as in Figure 21. The vertical direction (k fixed) corresponds to the p-version while the horizontal direction (r fixed) is nothing else but the traditional finite element approximation of fixed order on uniform grids. If we carry out our selection procedures as we did in 5.2 we will arrive at results for the h-version (we are still working within a horizontal line). Moreover, the theory of $A^s_{p,q}$ spaces is established for both horizontal and vertical directions (the latter is clear for $k = 0$, see the direct and inverse theorems for polynomial approximation in [DT], $k > 0$ needs some elaboration, compare also [Os13]). Now, the h-p-version might be viewed as the two-stage process of first selecting (for an individually given f or a class of functions) an arbitrary increasing path and then selecting subspaces within this path as we did for the h-version. As typical nodal basis in $V_{r,k}$, one should take the usual linear nodal basis functions ($r = 1$) complemented by the higher degree hiearchical bubble functions located in exactly one interval of

π_k^*. It is by no means clear at present whether this choice for $r > 1$ is a good one, and whether a selection procedure in vertical direction can be based on the same principles (basis function selection relying on the L_p-stability of the basis) that were used for the h-version (horizontal direction). Nevertheless, we may ask the following question. Given an arbitrary path in the two-dimensional scheme $\{V_{r,k}\}$, find an appropriate analog of the $A_{p,q}^s$ theory which allows for full or partial answers on the corresponding nonlinear approximation scheme based on basis function selection. Moreover, it is still interesting to know of some new examples of sequences of approximating subspaces (different from the vertical resp. horizontal lines) for which something interesting comes out. E.g., what about the diagonal?

It might seem that these questions are far away from being numerically relevant (as to us, this truly depends on the answers we can get). We will therefore look at a different theoretical interpretation of the h-p-version which is closer to the real adaptive algorithm. To this end, let us construct a tree-structured family of subspaces as in Figure 22 according to the following rules. To initialize the process, define as the root the subspace $V_{1,0} = S_1^0(\pi_0^*)$ (or any other low-dimensional $V_{r,k}$). Assume that after some steps a subspace V_α of $C[0,1]$ has been constructed which coincides with some $S_r^0(\pi)$ where π exclusively consists of dyadic intervals (of different length). The next step of the procedure will create all sons V_β of V_α which also describes the strucure of the mapping σ. To get a son, take any of the intervals of π and fix one of two possibilities: either subdivide this interval into two new dyadic intervals and preserve the polynomial degree of the old interval for both of them, or keep this interval but increase the polynomial degree by one. The new partition and the new degree vector create one of the potential sons! Clearly, such a construction leads to an explosion of sons when climbing further in the tree. But the construction itself is exactly what reasonable people would do in practice: take a local procedure of error estimation (a so-called error indicator) to decide whether to stop the refinement process at this interval because the error is below a certain level or to proceed with a) subdivision (h-version) or b) raising the polynomial degree (p-version). That one should keep this procedure local is obvious from the potential applications (to get a feeling about optimal distribution of intervals and degrees, we recommend looking at the detailed investigations and tests provided in [GB]). This process is sometimes called adaptive approximation based on an error indicator, and a common strategy is *equidistribution* of this error bound. Seen formally, the two-stage process in a tree structure (where the first stage consists in fixing recursively one son for each father (=setting the path), and the second in the local selection (=stopping criteria)) may lead to the same result as above in the two-parameter subspace scheme but the latter has a more restrictive understanding of paths.

Once again, one might ask stupid questions such as if there are function spaces behind the construction, and what are theoretically justified error estimators in this respect? Also, one has to get more insight in order to find new interesting situations intermediate to piecewise analyticity (lower dimensional singularity manifolds allowed) resulting in exponential rates, and finite smoothness (algebraic rates). Weighted spaces might be incorporated.

There are many problems not touched on in these notes, one of importance among them is the approximation theory of the p-version which corresponds to classical polynomial approximation. The latter has been thought (for many years) as a field not creating anything new for numerical analysis. But the contrary is true, look, for instance, at [BCM] where interesting questions have been fixed and solved.

5.4.2   Wavelet packets and compression

We rely on the papers by Wickerhauser et.al. [Wh, CMQ, CW] and another recent paper by Mallat, Zhang [MZ]. We only give an introduction to the subject which has been dealt with until now only in a pure Hilbert space fashion, and has something in common with the questions of this section.

For the following definitions, compare 2.1.3 and the textbooks [Db, Ch2]. Let $\{p_k\}$ be a bi-infinite sequence of real numbers with only finitely many non-zero terms, satisfying the properties

$$\sum_k p_{2k} = \sum_k p_{2k+1} = 2^{-1/2}$$

$$\sum_k p_k p_{k+2l} = 0 \qquad \forall \, l = 1, 2, \dots .$$

Examples are constructed in [Db], the simplest one

$$p_0 = p_1 = 2^{-1/2} \, , \quad p_k = 0 \quad \text{elsewhere}$$

yielding the Haar case. The mask $p \equiv \{p_k\}$ is called the summing filter, the conjugate (differencing) filter is introduced by $q_k = (-1)^k p_{1-k}$. Now define a hierarchy of wavelets on $\mathbf{R}^1$ by recursively applying the masks: first we define two basic functions w_0 and w_1 by the refinement equations

$$w_0(x) \;=\; 2^{1/2} \sum_k p_k w_0(2x - k)$$
$$w_1(x) \;=\; 2^{1/2} \sum_k q_k w_1(2x - k) \,.$$

Actually, these functions correspond to the ϕ and ψ of 2.1.3. After this, we create the higher w_n by repeating the following process

$$w_{2n}(x) \;=\; 2^{1/2} \sum_k p_k w_n(2x - k)$$
$$w_{2n+1}(x) \;=\; 2^{1/2} \sum_k q_k w_n(2x - k) \,.$$

For the Haar case mentioned above (where $w_0(x)$ is the characteristic function of $[0,1)$) the resulting system is the classical orthogonal Walsh system on $[0,1)$ (its periodized version models the trigonometric system in a discrete setting, for these classical systems you may consult [KS]). Finally, denote by

$$w_{n;j,i}(x) = 2^{j/2} w_n(2^j x - i) \,, \quad i,j \in \mathbf{Z}$$

the dilates/translates of the w_n, $n \in \mathbf{Z}_+$. The collection of all these functions is called wavelet packet (the reader may believe that due to the assumptions the $w_{n;j,i}$ are available through fast multilevel computations to any desirable accuracy).

This wavelet packet contains a lot of redundancy: several subsystems like the Haar-like wavelet system $\{w_{1;j,i}\}$ or the Walsh-like system $\{w_{n;0,i}\}$ build orthonormal bases of $L_2(\mathbf{R})$, and there are many other interesting subsystems. Since the different subsystems of wavelet packet functions represent qualitatively different Fourier-analytical behaviour (localization in time for $j \to \infty$ versus localization in frequency domain if $n \to \infty$) it is natural to use the redundancy to select, for a given signal $f \in L_2(\mathbf{R})$, an individual orthogonal subsystem reflecting its typical time-frequency properties. Discrete variants of this idea (i.e. wavelet packets of $\mathbf{R}^N$-vectors; N a power of two) have been used to design data compression algorithms for applications to speech analysis, see [Wh]. One simple idea is to check the decompositions of an arbitrarily given vector with respect to all possible basis subsets of the vectorial wavelet packet for possessing a minimum of large coefficients (above some bound ϵ). Once again, we may look at this procedure as a nonlinear method to find a subspace of minimal dimension by basis function selection from a certain subspace scheme, including relatively different approximation features (i.e. localization in time resp. in frequency domain). The interesting and unsolved approximation-theoretical problem is to find some characteristics of functions which describe the class of signals which can be represented more efficiently by these nonlinear wavelet packet compression algorithms in comparison to a compression working only. with the wavelet decomposition (with respect to $\{w_{1;j,i}\}$) or a pure Fourier-Walsh approach (decompositions with respect to $\{w_{n;0,i}\}$).

The analogous problem arises with a scheme proposed for adaptive time-frequency decomposition in [MZ]. In this paper, a continuous three-parameter family of atoms is introduced:

$$g_\gamma(t) = \frac{1}{\sqrt{s}} g\left(\frac{t-u}{s}\right) e^{i\xi t} , \qquad \gamma \equiv (s, \xi, u) ,$$

where $s > 0$ corresponds to the scale parameter j, $\xi \in \mathbf{R}$ to the frequency number n, and $u \in \mathbf{R}$ to the translation parameter i of the previous wavelet packet example. A simple energy analysis (see [MZ]) shows that in the time domain g_γ is concentrated in an interval around u with length proportional to s while in the frequency domain $\hat{g}_\gamma$ is localized in an interval around ξ of length proportional to $1/s$.

The algorithm proposed in [MZ] aims at obtaining for each signal $f(t) \in L_2(\mathbf{R})$ an individual best decomposition

$$f(t) = \sum_{n=-\infty}^{\infty} a_n g_{\gamma_n}(t) .$$

Some a priori fixed choices of the γ_n correspond to well-known methods in signal analysis. For fixed $s_0 > 0$, and appropriately small $u_0, \xi_0 > 0$, the choice of the set $\{(s_0, k\xi_0, mu_0)\}$ for $\{\gamma_n\}$ $(k, m \in \mathbf{Z})$ corresponds to the window Fourier transform technique. Fixing some ξ_0, taking $\xi = \xi_0/s$, and discretizing the scale parameter s with respect to the geometric progression s_0^k and the translation parameter u uniformly with stepsize $u_0 s_0^k$, one arrives at a typical wavelet scheme. The point is now to allow for more general choices. To this end, the following basic step has to be carried out repeatedly: For a fixed $0 < \alpha < 1$, choose γ_0 according to

$$|(f, g_{\gamma_0})| \geq \alpha \sup_\gamma |(f, g_\gamma)| \qquad \left((f, g) = \int_\mathbf{R} f(t) g(t)\, dt \right) ,$$

and define Rf by

$$f = (f, g_{\gamma_0}) g_{\gamma_0} + Rf .$$

Now repeat the process with the first residual Rf instead of f, and so on. The final result is an infinite decomposition of $P_V f$, the L_2 orthogonal projection of f onto V, the closed linear hull of the g_γ. If g is appropriately chosen, the latter coincides with L_2 and one has even representation of f. The theoretical and algorithmical features of this process called matching pursuits are discussed in [MZ] in great detail, also special choices of g and numerical tests are provided there. Once again, one is tempted to ask for more theoretical insight and quantitative characterizations of signal classes for which this nonlinear selection process is particularly efficient. One may try to make the question more definitive by embedding it into a subspace scheme as mentioned at the beginning of this section. Also, since the matching pursuits algorithm is not restricted to the above particular situation, one may look for its applicability to serve as adaptivity criteria within multilevel p.d.e. solvers. All these questions need further specification.

5.4.3 Approximation with long rectangles

We close with a short discussion of a further interesting model situation for studying the more complicated nonlinear approximation processes we intend to popularize by this subsection. The partial answers obtained for the h-version in 5.2 are actually results about a restricted h-version: we did not allow for general triangulations (or partitions) into $\leq n$ triangles (or rectangles etc.) but only for those composed of nondegenerating triangles. Long, thin triangles and the need to use them arise naturally in some problems of recent research interest, for instance, to resolve boundary layers for convection dominated diffusion problems [KR1, KR2] or for nonlinear conservation laws (for the latter, the question comes out when attempting to carry over recent one-dimensional results by DeVore, Lucier [DL1] to a two-dimensional setting). Since quantitative approximation (direct and inverse theorems) by splines or finite elements on triangulations with arbitrarily directed and shaped triangles is beyond the author's imagination, we suggest a much simpler but still delicate model problem: a two-directional approximation scheme involving arbitrary dyadic rectangles as underlying partition structure.

For simplicity, consider the subspaces V_{j_1,j_2} of piecewise bilinear functions with respect to the tensor-product rectangular partition of the unit square in $\mathbf{R}^2$ with stepsize 2^{-j_1} resp. 2^{-j_2} in the coordinate directions x_1 resp. x_2. These subspaces have already been used in 3.5. We denote the nodal basis functions of V_{j_1,j_2} by $N_{j_1,j_2;i}$, and ask (in analogy to 5.2) for adequate estimates of the quantities

$$\bar{e}_n(f)_X = \inf_{\mathcal{I}\,:\,|\mathcal{I}|\leq n} \left\| f - \sum_{(j_1,j_2;i)\in\mathcal{I}} c_{j_1,j_2;i} N_{j_1,j_2;i} \right\|_X .$$

This means that $\|\cdot\|_X$ should be some reasonable L_p or Besov-Sobolev norm, and that the description of the function classes $Y \subset X$, for which some rate of nonlinear approximation by the above scheme holds true, yields nontrivial and close to optimal sufficient conditions on f to be better approximated by the above scheme involving long rectangles with sides parallel to the coordinate axes compared to, say, the results given in 5.2. The resulting conditions should, in some sense, interpolate the results for the path

$$V_{0,0} \subset V_{1,1} \subset \ldots \subset V_{j,j} \subset \ldots$$

resp.

$$V_{0,0} \subset V_{1,0} \subset \ldots \subset V_{j,0} \subset \ldots$$

which can be obtained along the lines of 5.2 and which correspond to the restricted h-version for functions $f(x_1, x_2)$ resp. $f(x_1)$.

Clearly, the question may be reformulated in terms of the second (tree-like) scheme mentioned at the beginning of 5.4, in order to discuss it in a setting which is closer to an adaptive algorithm based on locally deciding whether to subdivide a given rectangle in x_1 or x_2 direction or to stop the refinement in this rectangle. We believe that this is the simplest model problem among the above ones. Since we did not promise any definitive results in this section but just a rough discussion of an interesting direction of future research, we will now stop.

References

1   Textbooks and monographs

[Ad]   A d a m s, R. A.: Sobolev spaces. Acad. Press, New York 1975.

[Ah]   A h i e z e r, N. I.: Lectures on approximation theory. Goztechizdat, Moskva 1947 (in Russian). Dt. Übers.: Vorlesungen über Approximationstheorie, 2. Aufl., Akademie-Verlag, Berlin 1967.

[Al]   A l t, H.: Lineare Funktionalanalysis. 2.Auflage, Springer, Berlin 1992.

[AB]   A x e l s s o n, O.; B a r k e r, V. A.: Finite element solution of boundary value problems. Theory and computation. Acad. Press, New York 1984.

[BGO]   B a b u s k a, I.; G a g o, J.; O l i v e i r a, E. R. de A.; Z i e n k i e w i c z, O. C. (eds.): Accuracy estimates and adaptive refinement in finite element computations. Int. Conf. Lisbon 1984, Wiley, Chichester 1986.

[Ba]   B a n k, R.: PLTMG: A software package for solving elliptic partial differential equations. User's guide 6.0. Frontiers in Applied Mathematics v. 7, SIAM Publ., Philadelphia 1990.

[BL]   B e r g h, J.; L ö f s t r ö m, J.: Interpolation spaces. An introduction. Springer, Berlin 1976.

[BIN]   B e s o v, O. V.; I l' i n, V. P.; N i k o l' s k i j, S. M.: Integral representations of functions and embedding theorems. Nauka, Moskva 1975 (in Russian).

[Bo1]   d e B o o r, C.: A practical guide to splines. Springer, Berlin 1978.

[Bo2]   d e B o o r, C.: Splinefunktionen. Birkhäuser, Basel 1990.

[BHR]   d e B o o r, C.; H ö l l i g, K.; R i e m e n s c h n e i d e r, S.: Box splines. Appl. Math. Sci. v. 98, Springer, New York 1993.

[Br1]   B r a e s s, D.: Finite Elemente. Springer, Berlin 1992.

[Br2]   B r a e s s, D.: Nonlinear approximation theory. Springer, Berlin 1986.

140   References

[BF]   B r e z z i, F.; F o r t i n, M.: Mixed and hybrid finite element methods. Springer, Berlin 1991.

[BB]   B u t z e r, P. L.; B e r e n s, H.: Semi-groups of operators and approximation. Springer, Berlin 1967.

[BN]   B u t z e r, P. L.; N e s s e l, R. J.: Fourier analysis and approximation. Birkhäuser, Basel 1971.

[BS]   B u t z e r, P. L.; S c h e r e r, K.: Approximationsprozesse und Interpolationsmethoden. Bibl. Inst. Mannheim, Zürich 1968.

[CO]   C a r e y, G. F.; O d e n, J. T.: Finite elements. Mathematical aspects, v. IV. Prentice-Hall, Englewood Cliffs, 1984.

[CDM]   C a v a r e t t a, A. S.; D a h m e n, W.; M i c c h e l l i C.: Stationary subdivision, Memoirs Amer. Math. Soc. 453, v. 93, AMS, Providence 1991.

[Ch1]   C h u i, C. K.: Multivariate splines. CBMS-NSF Reg. Conf. Series in Applied Math. 54, SIAM Publ., Philadelphia 1988.

[Ch2]   C h u i, C. K.: An introduction to wavelets. Acad. Press, Boston 1992.

[Ch3]   C h u i, C. K. (ed.): Wavelets: A tutorial in theory and applications. Acad. Press, Boston 1992.

[Ci]   C i a r l e t, C.: The finite element method for elliptic problems. North-Holland, Amsterdam 1978.

[Db]   D a u b e c h i e s, I.: Ten lectures on wavelets. CBMS-NSF Reg. Conf. Series in Appl. Math. 61, SIAM Publ., Philadelphia 1992.

[Dg]   D a u g e, M.: Elliptic boundary value problems on corner domains. Lect. Notes Math. 1341, Springer, Berlin 1988.

[DL]   D a u t r a y, R.; L i o n s, J.-L.: Mathematical analysis and numerical methods for science and technology. Vol. 1-6, Springer, Berlin 1990-1992.

[DT]   D i t z i a n, Z.; T o t i k, V.: Moduli of smoothness, Springer, Berlin 1987.

[Dy]   D' y a k o n o v, V. G.: Optimization of computational work – asymptotically optimal algorithms for elliptic equations. Nauka, Moskva 1989 (in Russian).

[EE]   E d m u n d s, D. E.; E v a n s, W. D.: Spectral theory and differential operators. Clarendon Press, Oxford 1987.

[FJW]   F r a z i e r, M.; J a w e r t h, B.; W e i s s, G.: Littlewood-Paley theory and the study of function spaces. CBMS-AMS Reg. Conf. Series, SIAM Publ., Philadelphia 1991.

[GiR]   G i r a u l t, V.; R a v i a r t, P. A.: Finite element methods for Navier-Stokes equations. Springer, Berlin 1986.

[Gv]   G r i s v a r d, P.: Elliptic problems in nonsmooth domains. Pitman Monographs v. 24, Longman Sci. & Techn., Harlow 1985.

[GR]   G r o ß m a n n, C.; R o o s, H.-G.: Numerik partieller Differentialgleichungen. Teubner, Stuttgart 1992.

[Ha1]   H a c k b u s c h, W.: Theorie und Numerik elliptischer Differentialgleichungen. Teubner, Stuttgart 1986. Engl. Transl.: Elliptic differential equations. Theory and numerical treatment. Springer, New York 1992.

[Ha2]   H a c k b u s c h, W.: Iterative Lösung großer schwachbesetzter Gleichungssysteme. Teubner, Stuttgart 1991. Engl. Transl.: Iterative solution of large sparse systems of equations. Springer, New York 1994.

[HL]   H o s c h e k, J.; L a s s e r, D.: Grundlagen der geometrischen Datenverarbeitung. Teubner, Stuttgart 1989.

[Jo]   J o h n s o n, C.: Numerical solutions of partial differential equations by the finite element method. Cambr. Univ. Press, Cambridge 1988.

[KS]   K a s h i n, B. S.; S a a k y a n, A. A.: Orthogonal series. Transl. Math. Monographs v. 75, AMS, Providence 1989.

[KF]   K o l m o g o r o v, A. N.; F o m i n, S. V.: Reelle Funktionen und Funktionalanalysis. VEB Dt. Verl. Wiss., Berlin 1975.

[KN]   K ř í ž e k, M.; N e i t t a a n m ä k i, P.: Finite element approximation of variational problems and applications. Pitman Monographs v. 90, Longman Sci. & Techn., Harlow 1990.

[KS]   K u f n e r, A.; S ä n d i g, A.-M.: Some applications of weighted Sobolev spaces. Teubner-Texte Math. 100, Teubner, Leipzig 1987.

[Li]   L i g h t, W. (ed.): Advances in Numerical Analysis. V. II. Wavelets, subdivision algorithms, and radial basis functions. Clarendon Press, Oxford 1992.

[LM]   L i o n s, J.-L.; M a g e n e s, E.: Non-homogeneous boundary value problems and applications. V. 1-3, Springer, Berlin 1973.

142 References

[LD]   Lorentz, G. G.; DeVore, R. A.:   Constructive approximation. Grundlehren, v. 303, Springer, Berlin 1993.

[Mk]   Marchuk, G. I.: Methods of numerical mathematics. Springer, New York 1982.

[Mz]   Maz'ya, V. G.: Sobolev spaces. Springer, Berlin 1985.

[Nč]   Nečas, J.: Les methodes directes en theorie des equations elliptiques. Academia, Prague, 1967.

[Ni]   Nikol'skij, S. M.: Approximation of functions of several variables and imbedding theorems. 2nd edition, Nauka, Moskva 1977 (in Russian). Engl. transl. of the first edition: Springer, Berlin 1975.

[Nb]   Nürnberger, G.: Approximation by spline functions. Springer, Berlin 1989.

[PP]   Petrushev, P. P.; Popov, V. A.: Rational approximation of real functions. Encycl. Math. Appl., Cambridge Univ. Press, Cambridge 1987.

[ST]   Schmeisser, H.-J.; Triebel, H.: Topics in Fourier analysis and function spaces. Geest & Portig, Leipzig 1987, Chichester, Wiley 1987.

[Su1]   Schumaker, L. L.: Spline functions : basic theory. Wiley, New York 1981.

[St]   Stein, E. M.: Singular integrals and differentiability properties of functions. Princeton Univ. Press, Princeton 1970.

[SW]   Stein, E. M.; Weiss, G.:   Introduction to Fourier analysis on Euclidean spaces. Princeton Univ. Press, Princeton 1971.

[Tl]   Temlyakov, V. N.: Approximation of periodic functions. Nova Science Publishers, Inc., New York 1993.

[TWM]   Thompson, J. F.; Warsi, Z.; Mastin, C. W.: Numerical grid generation. Foundations and applications. North Holland, New York, 1985.

[Ti]   Timan, A. F.: Theory of approximation of functions of a real variable. Fizmatgiz, Moskva 1960 (in Russian). Engl. transl.: 1963.

[Tr1]   Triebel, H.: Höhere Analysis, VEB Dt. Verl. Wiss., Berlin 1972, Verl. Harri Deutsch, Thun 1980.

[Tr2]   Triebel, H.: Interpolation theory, Function spaces, Differential operators. Dt. Verlag Wiss., Berlin 1978, North-Holland, Amsterdam, New York, Oxford 1978.

[Tr3]   T r i e b e l, H.: Theory of function spaces. Geest & Portig, Leipzig 1983, Birkhäuser, Basel 1983.

[Tr4]   T r i e b e l, H.: Theory of function spaces II, Birkhäuser, Basel 1992.

[Va]   V a r g a, R.: Functional analysis and approximation theory in numerical analysis. CBMS-NSF Reg. Conf. Ser. in Appl. Math. 61, SIAM Publ., Philadelphia 1971.

[We]   W e r s c h u l z, A. G.: The computational complexity of differential and integral equations. An information based approach. Oxford Univ. Press, Oxford 1991.

[ZM]   Z i e n k i e w i c z, O. C.; M o r g a n, K.: Finite elements and approximation. Wiley-Intersc., New York 1983.

2   Surveys

[AL]   A g o s h k o v, V. I.; L e b e d' e v, V. I.: Poincare- Steklov operators and domain decomposition methods for variational problems. In: Marcuk, G. I. (ed.): Computational Methods and Systems. V. 2, Moskva, Nauka 1985, 173-227 (Russian).

[Bb1]   B a b u s k a, I.: Advances in the p and h-p versions of the finite element method. A survey. In: Proc. Singapore Conf., ISNM 86, Birkhäuser, Basel 1988, 31-46.

[Bb2]   B a b u s k a, I.: The p and h-p versions of the finite element method: The state of the art. In: Finite elements. Theory and applications. Proc. ICASE Workshop Hampton 1986, (1988) 199-239.

[BA]   B a b u s k a, I.; A z i z, A. K.: Survey lectures on the mathematical foundation of the finite element method. In: Aziz, A. K. (ed.): The mathematical foundation of the finite element method with applications to partial differential equations, Acad. Press, New York 1972.

[BS]   B a b u s k a, I.; S u r i, M.: The p and h-p version of the finite element method. An overview. Comp. Meth. Appl. Mech. Engin. 80, 1990, 5-26.

[BKL]   B e s o v, O. V.; K u d r y a v z e v, L. D.; L i z o r k i n, P. I.; N i k o l' s k i j, S. M.: Investigations in the theory of spaces of differentiable functions of several variables. Proc. Steklov Inst. Math. 1, 1990, 73-139.

[BC1]   B e y l k i n, G.; C o i f m a n, R.; R o k h l i n, V.: Wavelets in numerical analysis. Report 1991.

[Bo2]   d e B o o r, C.: Splines as linear combinations of B-splines. A survey. In: Lorentz, G. G. et. al. (eds.): Approximation Theory II. Acad. Press, New York 1976, 1-47.

144 References

[Bo3]   d e B o o r, C.: Quasiinterpolants and approximation power of multivariate splines. In: Dahmen, W.; Gasca, M.; Micchelli, C.; (eds.): Computation of curves and surfaces. Kluwer, Dordrecht 1990, 314-345.

[Bo4]   d e B o o r, C.: Multivariate piecewise polynomials. Acta Numerica 1993, Cambr. Univ. Press, Cambridge 1993, 65-110.

[BFK]   B ö h m, W.; F a r i n, G.; K a h m a n n, J.: A survey of curve and surface methods in CAGD. CAGD 1, 1984, 1-60.

[DM1]   D a h m e n, W.; M i c c h e l l i, C.: Recent progress in multivariate splines. In: Chui, C. K.; Schumaker, L.; Ward, J.; (eds.): Approximation Theory IV. Acad. Press, New York 1983, 27-121.

[De1]   D e V o r e, R. A.: Degree of approximation. In: Lorentz, G. G.; et. al. (eds.): Approximation Theory II. Academic press, New York 1976, 117-162.

[De2]   D e V o r e, R. A.: Degree of nonlinear approximation. In: Chui, C. K.; et. al. (eds.): Approximation Theory VI. Acad. Press, New York 1989, 175-201.

[DL2]   D e V o r e, R. A.; L u c i e r, B.: Wavelets. Acta Numerica 1992, Cambr. Univ. Press, Cambridge 1992, 1-56.

[DW1]   D r y j a, M.; W i d l u n d, O.: Towards a unified theory of domain decomposition for elliptic problems. In: Chan, T.; Glowinski, R.; Periaux, J.; Widlund, O. (eds.): 3rd Int. Symp. on Domain Decomposition Methods for PDE. SIAM, Philadelphia 1990.

[Fa]   F a r i n, G.: Triangular Bernstein-Bezier patches. CAGD 3, 1986, 83-127.

[Hö1]   H ö l l i g, K.: Multivariate splines. In: Approximation Theory Proc. Symp. Appl. Math. 36 AMS Providence 1986, 103-127.

[Hö2]   H ö l l i g, K.: Box splines. In: Chui, C. K.; Schumaker, L. L. ; Ward, J. (eds.): Approximation Theory V. Academic Press, New York 1986, 71-95.

[JL]   J a f f a r d, S.; L a u r e n c o t, Ph.: Orthogonal wavelets, analysis of operators, and applications to numerical analysis. In: [Ch3], 543-601.

[Su2]   S c h u m a k e r, L. L.: On spaces of piecewise polynomials in two variables. In: Singh, S. P. et. al. (ed.): Approximation Theory and Spline Functions. Reidel, Dordrecht 1984, 151-197.

[Te]   T e l y a k o v s k i j, S. A.: Research in the theory of approximation of functions at the mathematical institute of the academy of sciences. Proc. Steklov Inst. Math. 1, 1990, 141-197.

[Tr5]   T r i e b e l, H.: Einige neuere Entwicklungen in der Theorie der Funktionenräume. Jber. d. Dt. Math.-Verein. 89, 1987, 149-178.

[Ve1]   V e r f ü r t h, R.: A review of a posteriori error estimation and adaptive mesh-refinement techniques. Lectures held at TU Magdeburg 1993. Preprint Univ. Zürich; short version: J. Comput. Appl. Math. (to appear).

[Wi]   W i d l u n d, O.: Some Schwarz methods for symmetric and nonsymmetric elliptic problems. Tech. Rep. 581, Courant Inst., New York Univ., September 1991.

[Xu1]   X u, J.: Iterative methods by space decomposition and subspace correction. SIAM Review 34, 1992, 581-613.

[Ys1]   Y s e r e n t a n t, H.: Old and new convergence proofs for multigrid methods. Acta Numerica 1993, Cambr. Univ. Press, Cambridge 1993, 285-326.

3   Research articles

[BGS]   B a b u s k a, I.; G u o, B.; S t e p h a n, E.: On the exponential convergence of the h-p version of the boundary element method on polygons. Math. Meth. Appl. Sci. 12, 1990, 413-427.

[BCM]   B a b u s k a, I.; C r a i g, A.; M a n d e l, J.; P i t k ä r a n t a, J.: Efficient preconditioning for the p-version finite element method in two dimensions. SIAM J. Numer. Anal. 28, 1991, 624-661.

[BG]   B a b u s k a, I.; G u o, B.: The regularity of the solution of elliptic problems with piecewise analytic data. Part 1: Boundary value problem for linear elliptic equations of second order. SIAM J. Math. Anal. 19, 1988, 172-203. Part 2: The trace spaces and applications to boundary value problems with non-homogeneous conditions. SIAM J. Math. Anal. 20, 1989, 763-781.

[BB]   B a i, D.; B r a n d t, A.: Local mesh refinement multilevel techniques. SIAM J. Sci. Stat. Comp. 8, 1987, 109-134.

[BSW]   B a n k, R. E.; S h e r m a n, R. H.; W e i s e r, A.: Refinement algorithms and data structures for regular local mesh refinement. In: Stepleman, R.; et. al. (eds.): Scientific Computing IMACS, North Holland, Amsterdam 1983, 3-17.

[BS]   B a n k, R. E.; S m i t h, R. K.: A posteriori error estimates based on hierarchical bases. SIAM J. Numer. Anal. 30, 1993, 921- 935.

[Bä]   B ä n s c h, E.: Local mesh refinement in two and three dimensions, IMPACT Comput. Sci. Engng. 3, 1991,181-191.

146 References

[Be] Besov,O.V.: Extensions of functions to the frontier, with preservation of differential-difference properties in L_p. Matem. Sbornik 66, 108, 1965, 80-96 (in Russian).

[Bl] Beylkin,G.: On the representation of operators in bases of compactly supported wavelets. SIAM J. Numer. Anal. 29, 1992, 1716-1740.

[BC2] Beylkin,G.;Coifman,R.;Rokhlin,V.: Fast wavelet transforms and numerical algorithms. Comm. on Pure and Applied Math. 44, 1991, 141-183.

[BI] Binev,P.G.;Ivanov,K.G.: On a representation of mixed finite differences. Serdica 11, 1985, 259-268.

[BM] Bjørstad,P.E.;Mandel,J.: On the spectra of orthogonal projections with applications to parallel computing. BIT 31, 1991, 76-88.

[BDR] deBoor,C.;DeVore,R.;Ron,A.: Approximation from shift-invariant subspaces of $L_2(R^d)$. Trans. Amer. Math. Soc. 341, 1994, 787-806.

[BJ] deBoor,C.;Jia,R.Q.: A sharp upper bound on the approximation order of smooth bivariate piecewise polynomial functions. J. Approx. Th. 72, 1993, 24-33.

[Bm] Bornemann,F.A.: An adaptive multilevel approach to parabolic equations. I: General theory and 1D-implementation. IMPACT Comput. Sci. Engrg. 2, 1990, 279-317; II: Variable order time discretization based on multiplicative error correction. IMPACT Comput. Sci. Engrg. 3, 1991, 93-122; III: 2D error estimation and multilevel preconditioning. IMPACT Comput. Sci. Engrg. 4, 1992, 1-45.

[BE1] Bornemann,F.A.;Erdmann,B.;Kornhuber,R.: Adaptive multilevel methods in 3 space dimensions. Int. J. Numer. Math. Engin. 36, 1993, 3187-3203.

[BE2] Bornemann,F.A.;Erdmann,B.;Kornhuber,R.: A posteriori error estimates for elliptic problems in two and three space dimensions. Preprint SC 93-29, ZIB Berlin 1993.

[BY] Bornemann,F.A.;Yserentant,H.: A basic norm equivalence for the theory of multilevel methods. Numer. Math. 64, 1993, 455-476.

[BH] Bramble,J.H.;Hilbert,S.R.: Estimation of linear functionals on Sobolev spaces with applications to Fourier transforms and spline interpolation. SIAM J. Numer. Anal. 7, 1970, 112-124.

[BL1] Bramble,J.H.;Leyk,Z.;Pasciak,J.E.: The analysis of multigrid algorithms for pseudo-differential operators of order minus one. Manuscript, Febr. 1992.

[BL2]   Bramble, J. H.; Leyk, Z.; Pasciak, J. E.: Iterative methods for non-symmetric and indefinite elliptic boundary value problems. Math. Comp. 60, 1993, 1-22.

[BP1]   Bramble, J. H.; Pasciak, J. E.: A preconditioning technique for indefinite systems resulting from mixed approximations of elliptic problems. Math. Comp. 50, 1988, 1-17.

[BP2]   Bramble, J. H.; Pasciak, J. E.: The analysis of smoothers for multigrid algorithms. Math. Comp. 58, 1992, 467-488.

[BP3]   Bramble, J. H.; Pasciak, J. E.: New estimates for multilevel methods including the V-cycle. Math. Comp. 60, 1993, 447-471.

[BPS]   Bramble, J. H.; Pasciak, J. E.; Schatz, A. H.: The construction of preconditioners for elliptic problems by substructuring I. Math. Comp. 47, 1986, 1093-1120.

[BW1]   Bramble, J. H.; Pasciak, J. E.; Wang, J.; Xu, J.: Convergence estimates for product iterative methods with applications to domain decomposition. Math. Comp. 57, 1991, 1-22.

[BW2]   Bramble, J. H.; Pasciak, J. E.; Wang, J.; Xu, J.:   Convergence estimates for multigrid algorithms without regularity assumptions. Math. Comp. 57, 1991, 23-45.

[BX1]   Bramble, J. H.; Pasciak, J. E.; Xu, J.: Parallel multilevel preconditioners. Math. Comp. 55, 1990, 1-22.

[BX2]   Bramble, J. H.; Pasciak, J. E.; Xu, J.: The analysis of multigrid algorithms with nonested spaces or noninherited quadratic forms. Math. Comp. 56, 1991, 1-34.

[BXu]   Bramble, J. H.; Xu, J.: Some estimates for a weighted L^2 projection. Math. Comp. 56, 1991, 463-476.

[BLu]   Brandt, A.; Lubrecht, A. A.: Multilevel matrix multiplication and fast solution of integral equations. J. Comp. Phys. 90, 1990, 348-370.

[Bn1]   Brenner, S.: An optimal order multigrid method for P1 nonconforming finite elements. Math. Comp. 52, 1989, 1-15.

[Bn2]   Brenner, S.: An optimal order nonconforming multigrid method for the biharmonic equation. SIAM J. Numer. Anal. 26, 1989, 1124-1138.

[Bn3]   Brenner, S.: A nonconforming multigrid method for the stationary Stokes equations. Math. Comp. 55, 1990, 411-437.

148 References

[Bn4]   B r e n n e r, S.: Multigrid methods for nonconforming finite elements. In: Mandel, J. et.al. (eds.): Proc. Fourth Copper Mountain Conf. on Multigrid Methods, SIAM, Philadelphia 1989, 54-65.

[Bn5]   B r e n n e r, S.: Two-level additive Schwarz preconditioners for nonconforming finite element methods. Preprint, Clarkson Univ., Potsdam NY 1993.

[BK]   B r u d n y i, Ju. A.; K r u g l y a k, N. Ya.: About a family of approximation spaces. Coll. of papers on the "Theory of functions of several real variables". Jaroslavl 1978, 15-42.

[Bd]   B r u d n y i, Ju. A.: A multidimensional analogue of a theorem of Whitney. Math. USSR Sbornik 11, 1970, 157-170.

[Bg]   B u n g a r t z, H.-J.: Dünne Gitter und deren Anwendung bei der adaptiven Lösung der dreidimensionalen Poisson-Gleichung . Dissertation, TU München 1992.

[Bu]   B u r e n k o v, V. I.: On a method of extending differentiable functions. Trudy Mat. Inst. Steklov. 140 (1976). Engl. transl.: Proc. Steklov Inst. Math. 1, 1979, 27-70.

[BG]   B u r e n k o v, V. I.; G o l'd m a n, M. L.: On the extension of functions of L_p. Trudy Mat. Inst. Steklov. 150, 1979, 31-51 (in Russian).

[CX]   C a i, X.-C.; X u, J.: A preconditioned GMRES method for nonsymmetric or indefinite problems. Math. Comp. 59, 1992, 313-319.

[Ca]   C a i, X.-C.: An optimal two-level overlapping domain decomposition method for elliptic problems in two and three dimensions. SIAM J. Sci. Comput. 14, 1993, 239-247.

[CW]   C a i, X.-C.; W i d l u n d, O.: Multiplicative Schwarz algorithms for some nonsymmetric and indefinite problems. SIAM J. Numer. Anal. 30, 1993, 936-952.

[CGP]   C a i, Z.; G o l d s t e i n, Ch. I.; P a s c i a k, J. E.: Multilevel iteration for mixed finite element systems with penalty. SIAM J. Sci. Comput. 14, 1993, 1072-1088.

[CL]   C h u i, C. K.; L a i, M.-J.: Multivariate vertex splines and finite elements. J. Approx. Theory 60, 1990, 245-343.

[Cs]   C i e s i e l s k i, Z.: Constructive function theory and spline systems. Studia Math. 53, 1975, 277-302.

[Cl]   C l e m e n t, P.: Approximation by finite element functions using local regularization. RAIRO Anal. Num. 2, 1975, 77-84.

References   149

[CMQ]   C o i f m a n, R. R.; M e y e r, Y.; Q u a k e, S.; W i c k e r h a u s e r, M. V.: Signal processing and compression with wave packets. In: Proc. Conf. on Wavelets, Marseil-
le 1989.

[CW]   C o i f m a n, R. R.; W i c k e r h a u s e r, M. V.: Best adapted wave packet bases. Preprint, Yale Univ. 1990.

[CDJ]   C o h e n, A.; D a u b e c h i e s, I.; J a w e r t h, B.; V i a l, P.:   Multiresolution analysis,
wavelets and fast algorithms on the interval. C. R. Acad. Sci. Paris 316, 1993,
417-421.

[Cw1]   C o w s a r, L. C.: Dual variable Schwarz methods for mixed finite elements. Report
TR93-09, Dep. Math. Sciences, Rice University, Houston 1993.

[Cw2]   C o w s a r, L. C.: Domain decomposition methods for nonconforming finite element
spaces of Lagrange-type. Report TR93-11 Dep. Math. Sciences, Rice University,
Houston 1993.

[CT]   C r o u z e i x, M.; T h o m e e, V.: The stability in L_p and W_p^1 of the L_2 projection
onto finite element spaces. Math. Comp. 48, 1987, 521-532.

[Dk1]   D a h l k e, S.; W e i n r e i c h, I.: Wavelet-Galerkin-Methods: An adapted biorthogonal wavelet basis. Constr. Approx. 9, 1993, 237-262.

[Dk2]   D a h l k e, S.; W e i n r e i c h, I.: Wavelet bases adapted to pseudo-differential operators. Preprint A-92-9, Fachbereich Mathematik, FU Berlin, 1992.

[Dh]   D a h m e n, W.: Decompositions of refinable spaces and applications to operator
equations. Preprint 83, Inst. Geom. Prakt. Math., RWTH Aachen, 1993.

[DDS]   D a h m e n, W.; D e V o r e, R.; S c h e r e r, K.: Multidimensional spline approximation. SIAM J. Num. Anal. 17, 1980, 380-402.

[DK]   D a h m e n, W.; K u n o t h, A.: Multilevel preconditioning. Numer. Math. 63, 1992,
315-344.

[DM2]   D a h m e n, W.; M i c c h e l l i, C.: On the approximation order from certain multivariate spline spaces. J. Austral. Math. Soc. Ser. B 26, 1984, 233-246.

[DP1]   D a h m e n, W.; P r ö s s d o r f, S., S c h n e i d e r, R.: Wavelet approximation methods
for pseudodifferential operators II: Matrix compression and fast solution. Advances
in Comp. Math. 1, 1993, 259-335.

[DP2]   D a h m e n, W.; P r ö s s d o r f, S., S c h n e i d e r, R.:   Multiscale methods for pseudo-
differential equations. In: Schumaker, L.L.; Webb, G. (eds.): Recent Advances in Wavelet Analysis, Acad. Press, New York 1994, 191-235.

150   References

[DOS]   D a h m e n, W.; O s w a l d, P.; S h i, X.-Q.: C^1-hierarchical bases. J. Comp. Appl. Math. (to appear).

[DLY]   D e u f l h a r d, P.; L e i n e n, P.; Y s e r e n t a n t, H.:   Concepts of an adaptive finite element code. IMPACT Comput. Sci. Engrg. 1, 1989, 3-35.

[DJ1]   D e V o r e, R. A.; J a w e r t h, B.; L u c i e r, B.:   Surface compression. CAGD 9, 1992, 219-239.

[DJ2]   D e V o r e, R. A.; J a w e r t h, B.; L u c i e r, B.:   Image compression through wavelet transform coding. IEEE Trans. Inform. Theory 38, 2, 1992, 719-746.

[DJP]   D e V o r e, R. A.; J a w e r t h, B.; P o p o v, V.:   Compression of wavelet decompositions. Amer. J. Math. 114, 1992, 737-785.

[DL1]   D e V o r e, R. A.; L u c i e r, B.:   High order regularity for conservation laws. Indiana Univ. Math. J. 39, 1990, 413-430.

[DP1]   D e V o r e, R. A.; P o p o v, V.:   Free multivariate splines. Constr. Approx. 3, 1987, 239-248.

[DP2]   D e V o r e, R.; P o p o v, V.:   Interpolation of Besov spaces. Trans. Amer. Math. Soc. 305, 1988, 397-414.

[DS]   D e V o r e, R.; S h a r p l e y, R. C.:   Besov spaces on domains in $\mathbf{R}^d$. Trans. Amer. Math. Soc. 335, 1993, 843-864.

[DY]   D e V o r e, R.; Y u, X. M.:   Degree of adaptive approximation. Math. Comp. 55, 1990, 625-635.

[Df1]   D ö r f l e r, W.:   Hierarchical bases for elliptic problems. Math. Comp. 58, 1992, 513-529.

[Df2]   D ö r f l e r, W.:   A note on the preconditioner of Bramble, Pasciak, and Xu. Preprint Inst. Angew. Math. Univ. Zürich, 1991.

[DW2]   D r y j a, M.; W i d l u n d, O.:   Multilevel additive methods for elliptic finite element problems. In: Hackbusch, W. (ed.): Parallel Algorithms for PDE. Proc. 6th GAMM Seminar, Kiel 1990, Vieweg, Braunschweig 1991.

[DW3]   D r y j a, M.; W i d l u n d, O.:   Additive Schwarz methods for elliptic finite element problems in three dimensions. In: Chan, T. F.; Keyes, D. E.; Meurant, G. A.; Scroggs, J. S.; Voigt, R. G. (eds.): 5th Conference on Domain Decomposition Methods for PDE. SIAM Publ., Philadelphia 1992, 3-18.

[DW4]   D r y j a, M.; W i d l u n d, O.: Schwarz methods of Neumann-Neumann type for three-dimensional elliptic finite element problems. Preprint, Courant Inst., New York Univ., March 1993.

[DNW]   D u r a n, R.; N o c h e t t o, R. H.; W a n g, J.: Sharp maximum norm error estimates for finite element approximations of the Stokes problem in 2- D. Math. Comp. 51, 1988, 491-506.

[EG]   E l m a n, H. C.; G u o, X.-Z.: Performance enhancements and parallel algorithms for two multilevel preconditioners. SIAM J. Sci. Comput. 14, 1993, 890-913.

[El]   E l s c h n e r, J.: On the exponential convergence of some boundary element methods for Laplace's equation in non-smooth domains. In: Costable, M.; Dauge, M.; Nicaise, S. (eds.): Proc. Conf. on Boundary Value Problems and Integral Equations on Non-Smooth Domains, Luminy 1993 (to appear).

[EHK]   E r d m a n n, B.; H o p p e, R. H. W.; K o r n h u b e r, R.: Adaptive multilevel methods for obstacle problems in three space dimensions. Preprint SC 93-8, ZIB, Berlin 1993.

[EW1]   E w i n g, R. E.; W a n g, J.: Analysis of multilevel decomposition methods for mixed finite element methods. Tech. Rep., Univ. Wyoming 1991. R.A.I.R.O. (to appear).

[EW2]   E w i n g, R. E.; W a n g, J.: The Schwarz algorithm and multilevel decomposition iterative techniques for mixed finite element methods. In: Chan, T. F.; Keyes, D. E.; Meurant, G. A.; Scroggs, J. S.; Voigt, R. G. (eds.): 5th Conference on Domain Decomposition Methods for PDE. SIAM Publ., Philadelphia 1992, 3-18.

[FJ]   F r a z i e r, M.; J a w e r t h, B.: A discrete transform and decompositions of distribution spaces. J. Funct. Analysis 93, 1990, 34-170.

[FR]   F r e h s e, J.; R a n n a c h e r, R.: Asymptotic L^∞ estimates for linear finite element approximations of quasilinear boundary value problems. SIAM J. Numer. Anal. 15, 1978, 418-431.

[GN]   G a s t a l d i, L.; N o c h e t t o, R. H.: Optimal L_∞ error estimates for nonconforming and mixed finite element methods of lowest order. Numer. Math. 50, 1987, 597-611.

[Gr1]   G r i e b e l, M.: A parallelizable and vectorizable multi–level algorithm on sparse grids. In: Hackbusch, W. (ed.): Parallel Algorithms for PDE, Proc. 6th GAMM Seminar Kiel. Vieweg, Braunschweig 1991, 58-69.

[Gr2]   G r i e b e l, M.: Parallel multigrid methods on sparse grids. In: Hackbusch, W.; Trottenberg, V. (eds.): Multigrid Methods III. Series of Numer. Math. 98, Birkhäuser, Basel 1991, 211-221.

152   References

[Gr3]   G r i e b e l, M.: Multilevel algorithms considered as iterative methods on indefinite systems. Report SFB 342/29/91A TU München, October 1991.

[Gr4]   G r i e b e l, M.:   Grid- and point-oriented multilevel algorithms. TUM-I9224, Inst. f. Informatik, TU München, 1992.

[GO1]   G r i e b e l, M.; O s w a l d, P.: On additive Schwarz preconditioners for sparse grid discretizations. Numer. Math. 66, 1994, 449-463.

[GO2]   G r i e b e l, M.; O s w a l d, P.: Remarks on the abstract theory of additive and multiplicative Schwarz algorithms. Numer. Math. (to appear).

[GSZ]   G r i e b e l, M.; S c h n e i d e r, M.; Z e n g e r, C.: A combination technique for the solution of sparse grid problems. In: de Groen, P.; Beauwens, R. (eds.): Proceedings of IMACS International Symposium on Iterative Methods in Linear Algebra. Brussels, April 2. - 4. 1991. Elsevier, North Holland, Amsterdam 1992.

[GB]   G u i, M.; B a b u s k a, I.: The h, p and h-p versions of the finite element method in 1 dimension. Part I: The error analysis of the p-version, Part II: The error analysis of the h and h-p versions. Part III: The adaptive h-p version. Numer. Math. 49, 1986, 577-612, 613-657, 659-683.

[Ha3]   H a c k b u s c h, W.: The frequency decomposition multi-grid method I. Application to anisotropic equations. Numer. Math. 56, 1989, 229-245.

[Ha4]   H a c k b u s c h, W.: The frequency decomposition multi-grid method II. Convergence analysis based on the additive Schwarz method. Numer. Math. 63, 1992, 433-453.

[HN]   H a c k b u s c h, W.; N o w a k, Z. P.: On the fast matrix multiplication in the boundary element method by panel clustering. Numer. Math. 54, 1989, 463-491.

[Hn]   H a n i s c h, M. R.: Multigrid preconditioning for the biharmonic Dirichlet problem. SIAM J. Numer. Anal. 30, 1993, 184-214.

[HH]   H i p t m a i r, R.; H o p p e, R. H. W.: Mixed finite element discretization of continuity equations arising in semiconductor simulation. TUM-M9302, Math. Inst., TU München, 1993.

[HK]   H o p p e, R. H. W.; K o r n h u b e r R.: Adaptive multi-level methods for obstacle problems. SIAM J. Numer.Anal. (1994, to appear).

[Hu]   H u d s o n, S. M.: Polynomial approximation in Sobolev spaces. Indiana Math. J. 39, 1990, 199-228.

[Ja]   Jaffard, S.: Wavelet methods for fast resolution of elliptic problems. SIAM J. Numer. Anal., 29, 1992, 965-986.

[JaM]   Jawerth, B.; Milman, M.: Wavelets and best approximation in Besov spaces. In: Cwikel, M.; Milman, M.; Rochberg, R. (eds.): Interpolation spaces and related topics, Israel Math. Conf. Proceedings, Bar Ilan Univ., Ramat-Gan 1992, 107-112.

[JaS]   Jawerth, B.; Sweldens, W.: Wavelet multiresolution analyses adapted to the fast solution of boundary value ordinary differential equations. In: Melson, N.D; Manteuffel, T.A; McCormick, S.F. (eds.): Proc. Sixth Copper Mountain Conf. on Multigrid Methods, NASA Conf. Pupl. 3224, 1993, 259-273.

[JL]   Jia, R. Q.; Lei, J.: Approximation by multiinteger translates of functions having global support. J. Approx. Th. 72, 1993, 2-23.

[JM]   Jia, R. Q.; Micchelli, C.: Using the refinement equation for the construction of prewavelets II: Powers of two. In: Laurent, P. J. et. al. (eds.): Curves and Surfaces. Academic Press, New York 1991.

[JS1]   Johnen, H.; Scherer, K.: Characterization of generalized Lipschitz classes by best approximation with splines. SIAM J. Num. Anal. 11, 1974, 283-304.

[JS2]   Johnen, H.; Scherer, K.: On the equivalence of the K-functional and moduli of continuity and some applications. In: Schempp, W.; Zeller, K. (eds.): Constructive theory of functions of several variables, Birkhäuser, Basel 1977, 119-140.

[Ko]   Kornhuber, R.: Monotone multigrid methods for elliptic variational problems I, II. Report SC 93-18, SC 93-19, ZIB, Berlin 1993.

[KR1]   Kornhuber, R.; Roitzsch, R.: Adaptive Finite-Element-Methoden für konvektionsdominierte Randwertprobleme bei partiellen Differentialgleichungen. TECFLAM-Seminar, Stuttgart 1988, 103-116; also: Report SC 88-9, ZIB, Berlin 1988.

[KR2]   Kornhuber, R.; Roitzsch, R.: On adaptive grid refinement in the presence of internal or boundary layers. Report SC 89-5, ZIB, Berlin 1989.

[KY]   Kornhuber, R.; Yserentant, H.: Multilevel methods for elliptic problems on domains not resolved by the coarse grid. In: Keyes, D.; Xu, J. (eds.): Proc. 7th Symp. on Domain Decomposition Methods, Penn State Univ. 1993 (to appear); also: Report SC 93-21, ZIB, Berlin 1993.

[KO]   Kotyczka, U.; Oswald, P.: On piecewise linear prewavelets in R^2. Constr. Approx. (submitted).

154   References

[Ku]   K u n o t h, A.: Multilevel preconditioning. Dissertation, Freie Universität Berlin, Verlag Shaker, Aachen 1994.

[Ln]   L e i n e n, P.: Ein schneller adaptiver Löser für elliptische Randwertprobleme auf Seriell- und Parallelrechnern. Diss., Univ. Dortmund 1990.

[MZ]   M a l l a t, S.; Z h a n g, Z.: Matching pursuits with time-frequency dictionaries. Preprint, Courant Inst. New York Univ. 1992.

[Ma1]   M a t h e w, T. P.: Schwarz alternating and iterative refinement methods for mixed formulations of elliptic problems. Part I: Algorithms and numerical results. Numer. Math. 65, 1993, 445-468.

[Ma2]   M a t h e w, T. P.: Schwarz alternating and iterative refinement methods for mixed formulations of elliptic problems. Part II: Convergence theory. Numer. Math. 65, 1993, 469-492.

[MNe]   M a t s o k i n, A. M.; N e p o m n y a s c h i k h, S. V.: Schwarz alternating method in a subspace. Sov. Math. (Izv. vuz.) 29, 1985, 78-84.

[Mo1]   M o n k, P.: A finite-element method for approximating the time-harmonic Maxwell equations. Numer. Math. 63, 1992, 243-261.

[Mo2]   M o n k, P.: An analysis of Nedelec's method for the spatial discretization of Maxwell's equations. J. Comp. Appl. Math. (to appear).

[Ne1]   N e p o m n y a s c h i k h, S. V.: Decomposition and fictitious domain methods for elliptic boundary value problems. In: Chan, T. F.; Keyes, D. E.; Meurant, G. A.; Scroggs, J. S.; Voigt, R. G. (eds.): 5th Conference on Domain Decomposition Methods for PDE. SIAM Publ., Philadelphia 1992, 62-72.

[Ne2]   N e p o m n y a s c h i k h, S. V.: Fictitious components and subdomain alternating methods. Sov. J. Numer. Anal. Math. Modelling 5, 1990, 53-68.

[Ne3]   N e p o m n y a s c h i k h, S. V.: Method of splitting into subspaces for solving elliptic boundary value problems in complex-form domains. Sov. J. Numer. Anal. Math. Modelling 6, 1991, 151-168.

[Ne4]   N e p o m n y a s c h i k h, S. V.: Mesh theorems on traces, normalization of function traces and their inversion. Sov. J. Numer. Anal. Math. Modelling 6, 1991, 1-25.

[Ne5]   N e p o m n y a s c h i k h, S. V.: Application of domain decomposition to elliptic problems with discontinuous coefficients. In: Glowinski, R. et. al. (eds.): Fourth Int. Symp. on Domain Decomposition, SIAM, Philadelphia 1991.

[On]   O n g, E.: Hierarchical basis preconditioners for second order elliptic problems in three dimensions, Ph. D. Thesis, CAM Report 89-31, Dept. of Math., UCLA 1990.

[Os1]   O s w a l d, P.: On function spaces related to finite element approximation theory. Z. Anal. Anwendungen 9, 1990, 43-64.

[Os2]   O s w a l d, P.: On the degree of nonlinear spline approximation in Besov–Sobolev spaces. J. Approx. Th. 61, 1990, 131-157.

[Os3]   O s w a l d, P.: Hierarchical conforming finite element methods for the biharmonic equation. SIAM J. Numer. Anal. 29, 1992, 1610-1625.

[Os4]   O s w a l d, P.: On discrete norm estimates related to multilevel preconditioners in the finite element method. In: Ivanov, K. G.; Petrushev, P.; Sendov, B. (eds.): Constructive Theory of Functions, Proc. Int. Conf. Varna 1991 , Bulg. Acad. Sci., Sofia 1992, 203-214.

[Os5]   O s w a l d, P.: On estimates for hierarchic basis representations of finite element functions. Forsch.-Erg. FSU Jena, N/89/16, 1989.

[Os6]   O s w a l d, P.: Spline approximation in the metric L_p, $0 < p < 1$. Math. Nachr. 94, 1980, 69-96 (in Russian).

[Os7]   O s w a l d, P.: On a hierarchical basis multilevel method with nonconforming P1 elements. Numer. Math. 62, 1992, 189-212.

[Os8]   O s w a l d, P.: Multilevel preconditioners for discretizations of the biharmonic equation by rectangular finite elements. JNLAA (submitted); Forsch.-Erg. FSU Jena, Math. 3, 1991.

[Os9]   O s w a l d, P.: Norm equivalencies and multilevel Schwarz preconditioning for variational problems. Forsch.-Erg. FSU Jena, Math. 1, 1992.

[Os10]   O s w a l d, P.: Stable splittings of Sobolev spaces and fast solution of variational problems. Forsch.-Erg. FSU Jena, Math. 5, 1992.

[Os11]   O s w a l d, P.: On a BPX–preconditioner for P1 elements. Computing 51, 1993, 125-133.

[Os12]   O s w a l d, P.: Two remarks on multilevel preconditioners. Forsch.-Erg. FSU Jena, Math. 1, 1991.

[Os13]   O s w a l d, P.: On estimates for one-dimensional spline approximation. In: Schmidt, J. W.; Späth, H. (eds.): Splines in Numerical Analysis. Math. Research, v. 52, Akad. Verl., Berlin 1989, 111-124.

156   References

[Os14]   O s w a l d , P.: On the convergence rate of SOR: A worst case estimate. Computing (to appear), Forsch.-Erg. FSU Jena, Math. 4, 1993.

[Os15]   O s w a l d , P.: Stable subspace splittings for Sobolev spaces and some domain decomposition algorithms. In: Keyes, D.; Xu, J. (eds.): Proc. 7th Symp. on Domain Decomposition Methods, Penn State Univ. 1993. Extended version: Forsch.-Erg. FSU Jena, Math. 7, 1993.

[Os16]   O s w a l d , P.: Preconditioners for nonconforming elements. Math. Comp. (submitted), Forsch.-Erg. FSU Jena, Math. 3, 1993.

[PS]   P e e t r e , J.; S p a r r , G.: Interpolation of normed abelian groups. Ann. Mat. Pura Appl. 92, 1972, 217-262.

[Pi]   P i e t s c h , A.: Approximation spaces (analogies between spaces of sequences, functions and operators). J. Approx. Th. 32, 1981, 115-134.

[RSc]   R a n n a c h e r , R.; S c o t t , R.: Some optimal error estimates for piecewise linear finite element approximations. Math. Comp. 158, 1982, 437-445.

[RWZ]   R i e d e r , A.; W e l l s , O. R.; Z h o u , X.: A wavelet approach to robust multilevel solvers for anisotropic elliptic problems. Manuscript, Comp. Math. Lab., Rice Univ., Houston 1993.

[Rd1]   R ü d e , U.: Fully adaptive multigrid methods. SIAM J. Numer. Anal. 30, 1993, 230-248.

[Rd2]   R ü d e , U.: On the V-cycle of the fully adaptive multigrid method. Bericht I-9215, Inst. f. Informatik TU München, May 1992.

[Rd3]   R ü d e , U.: Mathematical and computational techniques for multilevel adaptive methods. Frontiers in Appl. Math. 13, SIAM, Philadelphia 1993.

[Ro]   R o i t z s c h , R.: Kaskade user's manual. TR 89-4, Konrad-Zuse-Zentrum f. Informationstechnik, Berlin 1989.

[Sr1]   S c h e r e r , K.: Characterization of generalized Lipschitz classes by best approximation with splines. SIAM J. Num. Anal. 11, 1974, 283-304.

[Sr2]   S c h e r e r , K.: On optimal global error bounds obtained by scaled local error estimates. Numer. Math. 36, 1981, 257-277.

[Su3]   S c h u m a k e r , L. L.: On super splines and finite elements. SIAM J. Numer. Anal. 26, 1989, 997-1005.

[Sc]   S c o t t, R.: Optimal L_∞ estimates for the finite element method on irregular meshes. Math. Comp. 30, 1976, 681-697.

[Sd]   S h a i d u r o v, V. V.: New convergence estimates for the symmetric V-cycle. In: GAMM-Seminar on Multigrid-Methods, Gosen 1992, IAAS-Report 5, Berlin 1993, 81-100.

[Sp]   S h a r p l e y, R.: Cone conditions and the modulus of continuity. In: Sec. Edmonton Conf. on Approximation Theory, CMS Conf. Proc. 3, AMS, Providence, R. I. 1983.

[Sm]   S m i t h, B. F.: A domain decomposition algorithm for elliptic problems in three dimensions. Numer. Math. 60, 2, 1991, 219-234.

[SW]   S m i t h, B. F.; W i d l u n d, O.: A domain decomposition algorithm using a hierarchical basis. SIAM J. Sci. Stat. Comput. 11, 1990, 1212-1220.

[SF]   S u z u k i, T.; F u j i t a, H.: A remark on the L_∞ bounds of the Ritz operator associated with a finite element approximation. Numer. Math. 49, 1986, 529-544.

[TCK]   T o n g, C. H.; C h a n, T. F.; K u o, C. C. J.: A domain decomposition preconditioner based on a change to a multilevel nodal basis. SIAM J. Sci. Stat. Comput. 12, 1991, 1486-1495.

[TW]   T r i e b e l, H.; W i n k e l v o ß, H.: Intrinsic atomic characterizations of function spaces on domains. Manuscript, FSU Jena 1994.

[VW1]   V a s s i l e v s k i, P. S.; W a n g, J.: Multilevel iterative methods for mixed finite element discretizations of elliptic problems. Numer. Math. 63, 1992, 503-520.

[VW2]   V a s s i l e v s k i, P. S.; W a n g, J.: Multilevel methods for P1 nonconforming finite element methods for elliptic problems. Preprint 1993.

[Ve2]   V e r f ü r t h, R.: A posteriori error estimates for non-linear problems. Finite element discretizations of elliptic equations. Math. Comp. (to appear 1994).

[Wa1]   W a n g, J.: Convergence analysis of multigrid algorithms for nonselfadjoint and indefinite elliptic problems. SIAM J. Numer. Anal. 30, 1993, 275-285.

[Wa2]   W a n g, J.: Convergence analysis of Schwarz algorithm and multilevel decomposition iterative methods II: non-selfadjoint and indefinite elliptic problems. SIAM J. Numer. Anal. 30, 1993, 953-970.

[WT]   W e i s s m a n, S. L.; T a y l o r, R. T.: Mixed formulations for plate bending elements. Comp. Meth. Appl. Mech. Engin. 94, 1992, 391-427.

158   References

[Wh]   W i c k e r h a u s e r, M. V.: Acoustic signal compression with wavelet packets. In [Ch3], 679-700.

[Wi]   W i d l u n d, O.: Some Schwarz methods for symmetric and nonsymmetric elliptic problems. In: Chan, T. F.; Keyes, D. E.; Meurant, G. A.; Scroggs, J. S.; Voigt, R. G. (eds.): 5th Conference on Domain Decomposition Methods for PDE. SIAM Publ., Philadelphia 1992, 19-36.

[Xu2]   X u, J.: Convergence estimates for some multigrid algorithms. In: Chan, T. F.; Glowinski, R.; Periaux, J.; Widlund, O. B. (eds.): III. Int. Symp. on Domain Decomposition Methods for Partial Differential Equations. SIAM, Philadelphia 1990.

[Xu3]   X u, J.: Counterexamples concerning a weighted L^2 projection. Math. Comp. 57, 1991, 563-568.

[Xu4]   X u, J.: A new class of iterative methods for nonselfadjoint or indefinite problems. SIAM J. Numer. Anal. 29, 1992, 303-319.

[Ys2]   Y s e r e n t a n t, H.: On the multi-level splitting of finite element spaces. Numer. Math. 49, 1986, 379-412.

[Ys3]   Y s e r e n t a n t, H.: Two preconditioners based on the multi-level splitting of finite element spaces. Numer. Math. 58, 1990, 163-184.

[Ys4]   Y s e r e n t a n t, H.: Hierarchical bases in the numerical solution of parabolic problems. In: Deuflhard, P.; Enquist, B. (eds.): Large Scale Scientific Computing, Birkhäuser, Boston, Basel, Stuttgart 1987.

[Ze]   Z e n g e r, C.: Sparse grids. In: Hackbusch, W. (ed.): Parallel Algorithms for PDE, Proc. 6th GAMM Seminar Kiel, Vieweg, Braunschweig 1991, 241-251.

[Zh1]   Z h a n g, X.: Multilevel Schwarz methods. Numer. Math. 63, 1992, 521-539.

[Zh2]   Z h a n g, X.: Studies in domain decomposition: Multilevel methods and the biharmonic Dirichlet problem. Ph. D. Thesis, Tech. Rep. 584, Courant Inst., New York Univ., September 1991.

[ZZ]   Z i e n k i e w i c z, O. C.; Z h u, J. Z.: A simple error estimator and adaptive procedure for practical engineering analysis. Int. J. Numer. Meth. Eng. 24, 1987, 337-357.

Index